Chicago Public Library

C
P
L

REFERENCE

D1249310

THE SUN
AND THE AMATEUR ASTRONOMER

THE SUN
AND THE
AMATEUR ASTRONOMER

by

W. M. BAXTER

Drawings by the author

DRAKE PUBLISHERS INC NEW YORK

LCCCN 72-7516
ISBN 0-87749-370-7

Published in 1973 by
Drake Publishers Inc
381 Park Avenue South
New York, N.Y. 10016

Printed in Great Britain

To

PAULINE and GERARD

CONTENTS

LIST OF DRAWINGS

LIST OF PLATES

ACKNOWLEDGEMENTS

THE AUTHOR WISHES TO ACKNOWLEDGE the source of the following photographs: Pls. I and II—Mount Wilson Observatory; Pl. III—H. C. Hunt; Pl. IV—Hans Arber; Pl. V—Paris Observatory (Meudon); Pl. VI—Climax Observatory; Pl. VIII—H. N. D. Wright; Pls. IX and X*a* —Patrick Moore.

The remaining photographs were taken by the author.

INTRODUCTION

THERE ARE MANY GOOD BOOKS on the Sun, but, on the whole, they are technical books describing the equipment, discoveries and work of the large professional observatories—some of which specialize in solar work—and they do not cater for the amateur astronomer with his limited facilities. Other books touch on the Sun, but only as one of the very many celestial bodies of interest to those of us who treat astronomy as a pleasing and satisfying hobby. There remains a gap, therefore, in the literature on the Sun which the present book endeavours to fill. It gives only general information likely to be of interest to the amateur observer, with an account of the author's work and that of others in this field, all with modest equipment.

Although amateur astronomers, and others interested in the heavenly bodies, will be familiar with the main features of the universe from such works as Patrick Moore's *The Amateur Astronomer* and from general books on astronomy, it is important that before the study of any particular object is undertaken its relationship to the other heavenly bodies, and in fact to the universe as a whole, should be fully appreciated. The first chapter deals with this aspect of the Sun at the risk of repeating information that is well known and elementary, but the knowledgeable amateur can at least commence reading from Chapter Two if he so wishes.

I started observing, as so many amateurs do, as a small boy with a home-made cardboard tube telescope, containing an "objective" consisting of a simple monocle, obtained from the local optician, with a small hand-magnifier for an eyepiece. Later, I obtained a 3-inch achromatic telescope mounted as an altazimuth on a tripod stand and I soon appreciated the improved images. I became a Member of the British Astronomical Association in 1932 and the friendly advice I obtained from my fellow members encouraged me to higher aspirations, resulting in the purchase, second-hand, of my present fine Cooke 4-inch achromatic refractor described in Chapter Six.

From the earliest I studied general books on astronomy (although there was nothing like the large number that is available today), and I am afraid I looked through my telescope at all and sundry. Having a general interest in astronomy, observing the heavens in all its aspects and showing your friends the moon and the planets, is very pleasureable, but I now appreciate that the greatest interest comes from making a special study, whichever object is chosen. Some years ago I chose the Sun, and my only regret is that I did not decide to specialize on that splendid object much earlier. Given clear skies, the Sun is at least visible every day which is more than can be said of all the other heavenly bodies and, moreover, there is no lack of visibility due to faintness of its light.

This book will seem very elementary to the solar experts, but my aim has been to encourage other amateur astronomers who, maybe, have only studied "the sky at night", and perhaps others, too, who have only a general interest in astronomy, to experience the pleasure of observing our nearest star—the Sun. We may know a lot, but there is far more that remains unexplained, and the more we learn the more there is to learn. The ever-changing face of the Sun is a constant source of interest, and watching and recording its periodic variations is both satisfying and rewarding.

My thanks are due to the amateur solar observers mentioned in the text, to many others not mentioned and in particular to Patrick Moore for his encouragement, without which the book would not have been written.

FOREWORD

SOLAR RESEARCH IS ONE of the most important branches of modern astronomy. Great attention is paid to it at many professional observatories, and elaborate special equipment has been designed. Daily records are made, and there can be few disturbances on the Sun which escape notice.

Under the circumstances it might seem that the amateur astronomer, working with modest equipment, would be unable to contribute anything of value, and would be confined to looking at sunspots and taking pictures of them for his own amusement. This is not, however, a correct summary of the situation, and in solar work, as in other branches of astronomy, the amateur can still make himself very useful —as well as enjoying himself and gaining a tremendous amount of knowledge in the process.

One such amateur is W. M. Baxter, Secretary of the British Astronomical Association, whose observatory at Acton is devoted almost entirely to the Sun. He has an international reputation as a solar observer, and the quality of his work can be judged from the photographs given in this book, but he would be (and is!) the first to point out that he is no professional scientist. Yet in astronomical circles, the name of "Baxter" is immediately associated with "the Sun".

There have been many books devoted to solar studies. Some are theoretical; some deal with practical observation from the professional viewpoint; and nearly all are excellent. However, there has never before been a book devoted to amateur observation of the Sun, and the present volume fills a real gap. In Chapter One the author has given a brief introduction and has then described how observations of the Sun are made with amateur equipment. His great experience enables him to point out all the pitfalls which await the novice, and anyone who is interested in the Sun and wants to take up serious observation can do no better than to read what he has written. Much of the material will not be found elsewhere, either in books or papers.

I believe that there will be many people who will enjoy reading the book as much as I myself did when I first saw it in manuscript form, and I believe, too, that it will play its part in encouraging many amateurs to begin studying the Sun.

PATRICK MOORE

FOREWORD TO THE

SECOND EDITION

THE DEATH OF W. M. Baxter in late 1971 came as a sad blow to his many friends in England and abroad. Over the years he had built up a unique reputation; his photographs of the Sun were known (and were used) everywhere, both by amateurs and often by professional astronomers also. But more than this, he was regarded—rightly—as an example of the very best kind of amateur astronomer; enthusiastic in his subject, skilful, and always ready to help and encourage others. He is badly missed.

For this new edition of his book, I have made the minimum corrections in order to bring it up to date. In fact not much needed to be done, since the information is basic, and the photographs pay testimony to Bill Baxter's skill.

PATRICK MOORE

Chapter One

THE SUN IN SPACE

BEFORE WE STUDY the Sun itself it will be useful to consider its relationship to other suns—the stars—and to the universe as a whole. As mentioned in the Introduction, our Sun is a star, much like other stars; and being relatively near to us it shows us a disk of light, whereas all the other stars, even the nearest ones, are so very far away that they appear as just points of light even in our largest telescopes. Hence the Sun, besides being well worth studying for itself, gives us a good idea of what other stars are like, as we are really having a close-up view of a typical star—"close-up" of a mere 93 million miles, for this is the distance the Sun is away from us. This distance, although seemingly large, is quite small compared with the distances of the other heavenly bodies apart from the members of our "local" solar system. In dealing with astronomical subjects we naturally have to get used to "astronomical" figures!

When we see photographs of the starry heavens, we see them as flat pictures, of two dimensions only, and we see the same effect when studying the actual night sky, as all the objects are much too far away for us to obtain the stereoscopic effect by which we judge relative distances of close objects. The ancient observers, in fact, thought that all the stars were fixed on a crystal sphere surrounding the earth and revolving, with the earth at its centre, once a day, giving the effect of their daily rising and setting.

It is quite an interesting and instructive experiment to make one's own stereoscopic pictures of a well-known constellation, such as Orion, by making a double sketch of the star positions side by side on black paper—the two pictures being the distance of the eyes apart—then pricking holes of different sizes (corresponding to the different brightnesses of the stars) in the correct positions on one picture, but on the other making the pricks very slightly displaced horizontally, some to the right and some to the left. Then view the pair against the light through a simple stereoscope, and a fascinating 3-D picture of Orion is seen. Let me hasten to add that it will not show the true relative positions of the stars, but it will give the feeling of three-dimensional space and help one to lose the feeling one always has of stars all appearing to be in one plane.

Another way to think of three-dimensional space is to imagine the room one is in to be full of bright points of light scattered all around, some in front and some behind. In fact if one imagines this mass of "stars" as being gathered into the form of a large cartwheel laid flat, one has a very good picture of our galaxy—that is our own particular mass of stars containing the Sun with its planets, including the Earth.

If one then looks through this mass of stars across a diameter of the wheel, the points of light will be seen in large numbers, whereas if one looks up or down, through the width of the wheel, the points of light will appear sparse. This is the effect we see in the sky on a clear, but dark, moonless night. The Milky Way shines as a bright band of closely packed stars spread across the sky—the diameter of our wheel—whereas away from the band the stars will be relatively sparse. The wheel of stars, the Milky Way, is our own galaxy and contains all the stars we see with the naked eye. The universe contains millions of such separate galaxies, each a mass of stars, and our large telescopes look so far into space that there is an unending supply of these objects to be seen and photographed.

Plate I (*a*) shows a typical area of the sky where the Milky Way is very rich in stars.

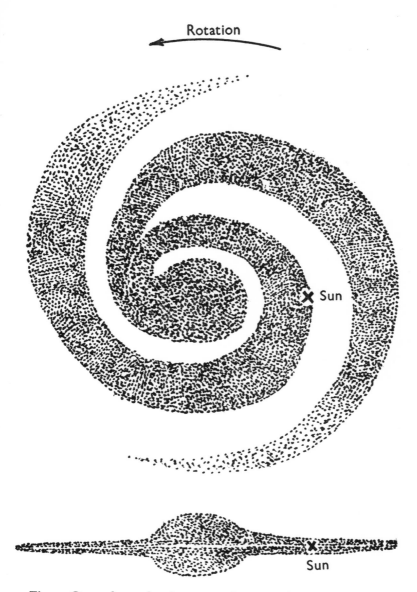

Fig. 1. Our galaxy, showing approximate position of the Sun.

Plate I (*b*) shows the nearest separate galaxy similar to our own, known as the Andromeda nebula. This is not really a nebula (all gas) but is a spiral galaxy containing its own vast quantity of stars, gas and dust, just as our own galaxy does.

Plate II (*a*) shows another of these gigantic spiral galaxies, that known as M 81 in the constellation Ursa Major or the Great Bear; it is flat-on as seen from the earth. Plate II (*b*) is a similar galaxy, but one which is seen edge-on from the Earth. It is in the constellation Coma Berenices, and is catalogued as N.G.C. 4565. All these depicted are very similar in shape to our own galaxy and, to put it more crudely, are rather like a fried egg being more or less flat, circular, and with a swelling in the centre. With their spiral arms they also rather remind one of a Catherine-wheel. Notice the band of obscuring gas and dust across the edge-on galaxy in Plate II (*b*). Yes, the universe contains not only stars in abundance, but a great deal of gas—some bright, some dark—and interstellar dust.

Coming back to our neighbourhood, Fig. 1 is a sketch of our galaxy as it would appear if seen from a distance both flat-on and edge-on, with the approximate position of our own star, the Sun, indicated. This star happens to have a retinue of planets going round it, of which our Earth is one, and we are certainly not in the centre of things! This huge spiral galaxy of ours is rotating, with the stars nearest the centre going round faster than those nearer the outside, in the same way that the planets nearer the Sun revolve round that body faster than the outer planets. It is estimated that the Sun makes one circuit of the galaxy in about 200 million years. This seems a long time; but as the age of the Earth is considered to be (on scientific evidence) of the order of 4,700 million years, it must have accompanied the Sun on quite a number of circuits already.

Although the universe contains millions of millions of

stars (each a huge sun) and also vast quantities of interstellar gas and dust, space is nevertheless very empty. Distances are great compared with the dimensions of stars,
huge as these are, and to visualize the emptiness let us consider a model. Imagine the Sun to be represented by a table-
tennis ball situated in London. The Earth by comparison
would be a tiny seed (only 0·014 inch diameter) circling
the ball at a distance of about 13 feet. But the *nearest* star
would be another table-tennis ball, to the North at about
Oslo or to the South at Barcelona! Or, in the U.S.A., one
ball would be in New York and the next in Chicago. Certainly
the balls (stars) scattered around at these distances apart
would seem very sparse and the intervening space very
empty. And yet we know there are hundreds of millions of
galaxies, each containing thousands of millions of stars. It
makes one feel very humble, to say the least.

Our Sun is an average star; some are smaller, some
hotter, while some are very much larger and some cooler.
The Earth makes its annual journey round the Sun at a
distance of 93 million miles; this is the average distance, as
the path is slightly elliptical, as are the paths of all the
planets. In fact we are about 3 million miles nearer the Sun
in our northern hemisphere winter than in our summer;
not that this comforts us on our chilly days, as the difference
is so very slight and is "swamped" by the difference due to
the seasons.

The Earth's axis, about which it makes its daily rotation,
is not perpendicular to the plane of the ecliptic (the Earth's
path round the Sun), but is inclined at an angle of $23\frac{1}{2}$
degrees. The axis maintains its direction in space as the
Earth goes round the Sun so that first the northern end of
the axis is turned Sunwards and then the southern end.
This gives us our seasons. When the northern end of the
axis tilts towards the Sun, the northern hemisphere of the
Earth has longer days and shorter nights and the Sun rises

to a higher elevation midday (summer), while at the same time the southern hemisphere experiences a lower midday sun, shorter days and longer nights (winter). The opposite applies, of course, when the southern end of the Earth's axis tilts towards the Sun; the southern hemisphere of the Earth has summer and the northern hemisphere has winter.

This will be clear from Fig. 2 (*a*) and (*b*). These drawings also show why in northern mid-summer the inhabitants of the Arctic Circle have 24 hours of daylight, while those in the Antarctic Circle, at the same time, are putting up with 24 hours of night. Six months later the Earth has completed half its annual journey and conditions are reversed. Then it is the turn of the Arctic dwellers to suffer the prolonged darkness, while those in the Antarctic experience long hours of sunshine.

In the northern mid-summer, people living along the latitude of the Tropic of Cancer have the Sun exactly overhead at midday while this applies to those on the latitude of the Tropic of Capricorn in their southern hemisphere mid-summer. Between these latitudes, that is to say "in the tropics", people have the Sun overhead at midday at least sometime during the year. In latitudes further north and south the Sun is never overhead.

The solar observer can see the Sun on every clear day, but its height in the sky depends on his latitude on the Earth's surface as well as on the time of year and time of day. In summer the Sun rises high in the sky and we feel maximum heat, whereas in winter the Sun remains low and the same area (or "cross-section") of its rays comes at a low angle; the rays are spread over a much wider area of the Earth, so that they feel relatively feeble to us in any particular spot.

Fig. 3 shows the height of the Sun above the horizon, in summer and in winter, as viewed from London.

24

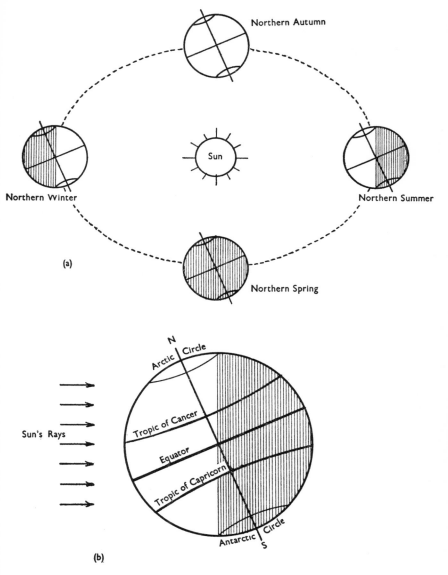

Fig. 2. The seasons.

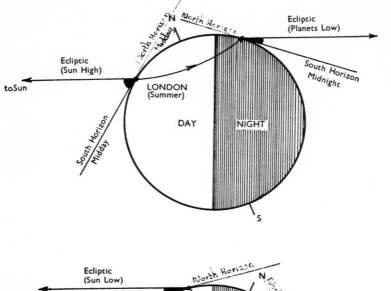

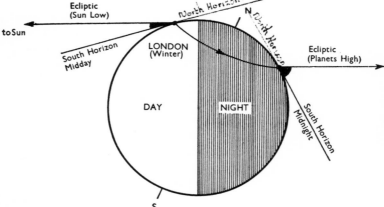

Fig. 3. Height of Sun as seen midday in London in summer and in winter.

The paths of the planets are nearly in one plane, and the moon's path round the Earth is nearly in that plane—the ecliptic, so called as eclipses of the Sun and of the Moon can only take place when those bodies are exactly in that plane. Sometimes the Moon passes between us and the Sun, and if its path crosses the Sun *exactly* we see a total solar eclipse. By a curious coincidence our small Moon (about a quarter of the Earth in diameter) is so near to us, only 240,000 miles away, that its apparent diameter in the sky is just about the same as the apparent diameter of the much larger Sun, 864,000 miles in diameter, 93 million miles away. The Moon, for a few minutes, can completely cut off the light of the Sun, casting a shadow patch on the Earth. The people in this small area of not more than 170 miles wide, which drifts across the surface of the Earth as the Moon passes the Sun, see the Sun blotted out and the sky darken. This phenomenon can only happen at the time of a new moon, of course, when the Moon is exactly in line with the Sun and between us and the Sun. The last total eclipse visible in Europe was seen in England as a partial eclipse, but the centre of totality passed across southern France, central Italy, Yugoslavia and South Russia. This is mentioned again in Chapter Five. See also Plate III.

Outside the central path of the Moon's shadow a partial eclipse of the Sun is seen, the proportion of the Sun's disk covered depending on the observer's distance from the central path. This partly shadowed area is about 2,000 miles wide.

Plate XII (top) is a photograph taken of the eclipse of 1959, October 2, as seen from London.

Sometimes when the Moon passes centrally in front of the Sun its diameter does not quite cover the Sun's full disk (when the eclipse takes place at a time when the Moon is furthest from us in its orbit) and on these occasions we do not observe a total blotting out of the Sun, as a small ring of sunlight is seen surrounding the slightly smaller black disk of the Moon. This is known as an "annular" (ring) eclipse of the Sun.

When the Moon is full, and therefore on the opposite

side of us to the Sun, we sometimes have a lunar eclipse when the Moon is darkened as it passes through the shadow of the Earth projecting into space. This phenomenon, while of general interest, is not of so much importance to astronomers as a solar eclipse, for reasons that will be made clear in later chapters, so we will leave the matter of eclipses for the moment. I would like to add one point, however. The question is sometimes raised why the Moon does not always eclipse the Sun when it "goes by" on its monthly journey, that is to say at every new moon.

Let us think back to our analogy of the tiny seed 0·014 inch diameter going round a table-tennis ball at a distance of 13 feet and add, in our imagination, an even tinier speck (about 0·0035 inch diameter or 3½ thousandths of an inch) representing the Moon and going round the seed about ½ inch away. As the Moon's path round the Earth is inclined to the path of the Earth round the Sun, the shadow of the speck (cast by light from the ball) could not be expected to fall *exactly* on the seed every time it came between seed and ball. The three bodies would have to be exactly in line for eclipses to happen, and this certainly does not occur every month.

In considering astronomical distances we need a handier scale than miles, and it is customary to relate distances to the time light takes to make the journey. Light travels at the stupendous speed of 186,000 miles a second. It takes 8½ minutes to reach us from the sun, 93 million miles away, and only 1½ seconds from our nearest neighbour, the Moon. But light takes over four years to reach us from the nearest star, and from the nearest spiral galaxy—that in the constellation Andromeda, Plate I (*b*)—2 million years. Our galaxy (Fig. 1) is about 100,000 light years in diameter, with the Sun about 30,000 light years from the centre, but it is only about 5,000 light years thick. The stars forming the constellations are all in our galaxy and relatively near the Sun; but although they are all speeding along in space, they are so far away that their configurations seem hardly changed during the thousands of years that men have watched them.

The most distant galaxies photographed by our largest telescopes are so far away that their light has taken something like 5,000 million years to reach us; we see them not as they are now but as they were 5,000 million years ago. In the same way we see the Sun, at any instant, as it was $8\frac{1}{2}$ minutes previously.

Another way to measure vast distances is in "parsecs", one parsec being defined as the distance a star would have to be for the angle subtended by the radius of the Earth's orbit, as seen from the star, to be 1 second of arc. One parsec equals 3·26 light years.

The stars are classified by magnitudes. A few are brighter than magnitude 1, while with the naked eye it is possible to see the fainter stars down to about magnitude 6. The higher the number, the fainter the star. Stars are also sometimes classified by "absolute magnitudes". These are their actual relative brightnesses independent of distance —as if all were at the same distance from us. The "standard distance" is taken as 10 parsecs (32·6 light years) and if our Sun were placed at this distance from us it would appear as a mere 5th magnitude star, just visible to the naked eye, while many of the other stars would be far brighter.

All the stars, and this goes for our Sun too, are self-luminous, being what one might call huge balls of fiery gas, but the planets circling the Sun, and their attendant moons, shine only by the sunlight reflected to us from their surfaces. Even the artificial satellites sent up by man shine like stars only when they catch the light of the Sun and disappear from sight when they pass into the Earth's shadow, as they do from time to time. If the Sun were to be blotted out all the other stars would continue to shine, but the planets and their satellites (natural and artificial) would disappear from view.

The Sun is all-important to us. Our very lives depend on its light and warmth. Some of the other stars may have planets going round them; we believe so, although they are much too far away for us to detect any such bodies, but there is every reason to think that amongst the many millions of stars there must be some with planets, and some of the latter in a similar physical state to our own Earth. We are left wondering if there are other beings not unlike ourselves elsewhere in the universe. Why not? We can only speculate at this stage.

We do know that our Sun has nine planets circling round it, one being our Earth with its attendant moon. Moreover, we are just at the right distance from the Sun for life, as we know it, to exist and flourish. If we were nearer we would scorch up, while if we were further away we would perish with the cold. Our food, our clothing, our power (coal, oil and water), our seasons and the rain, all originate from solar radiation, and although the Sun is continuously radiating energy at a colossal rate its mass is so very great that we can console ourselves with the knowledge that it can continue to supply our needs for millions of years to come. The Sun is pouring out this energy in all directions and, when you consider our earlier analogy of the table-tennis ball and the seed going round it 13 feet away, it will be appreciated what a minute fraction of the Sun's total radiation is captured by the Earth.

The Sun has gravitational control over all its family of planets and is moving in space—towards that part of the sky containing the constellation Hercules—at 12 miles a second. As we are carried along with it, at the same time as we are revolving round it once a year (at a speed of about 18½ miles a second) while spinning round on the Earth once a day, it will be seen that our movements in space, as individuals, are very complicated indeed.

We have now visualized the Sun's position in space relative to the universe, to our own galaxy, and to our earth-moon system. I realize that in this first chapter I have been telling amateur astronomers what most of them already know, but I felt it to be important to set the scene, as it were—or should I say the back-cloth?—and we can now proceed to the more detailed study of our subject—the Sun.

Chapter Two

PHYSICAL DETAILS

IN THE PRECEDING CHAPTER the Sun and stars were likened to huge balls of fiery gas. They are gaseous and our Sun is certainly huge and at a very high temperature. It is 864,000 miles in diameter, or just over 100 times the diameter of the Earth. We see it in the sky as a whitish-yellowish disk, or even red when setting or seen through a fog, but this is due to the effects of our atmosphere as its light is really white; and white being a mixture of all colours, sunlight is responsible for the full range of beautiful colours we see all around us—the sky in its varied tints, the colours of the flowers, of pictures, of paints and so on.

The stars themselves vary in colour, and this is an indication of their temperatures. The surface temperature of the Sun can be measured, and is found to be about 6,000 degrees C. Its mass is large and the pull of gravity at its surface is 28 times that on the surface of the Earth, so that a 1 lb weight here would weigh 28 lbs on the Sun. Its average density, however, is low compared with that of the Earth —about one quarter—being only $1\frac{1}{2}$ times the density of water. The density near the surface is *extremely* low, but it is very dense indeed towards the centre. The Sun's temperature, too, increases enormously towards the centre and it is estimated to be of the order of 14 million degrees.

As seen from the Earth, the Sun's disk subtends an angle of just over $\frac{1}{2}$ degree, being about the same as that subtended by the Moon, as mentioned in Chapter One. Actually, the Sun is so large that its diameter is roughly double that of the Moon's path round the Earth.

The Sun is continuously pouring out a vast amount of

energy into space, and thereby losing mass at a great rate. It is estimated that some 4 million tons of solar matter are being converted into energy every second, but the Sun is so huge that it can continue to do this for millions of years with no appreciable effect on its size or mass. There is no doubt that this vast output of energy is maintained internally by nuclear processes, rather like the continuous release of energy from millions of gigantic thermonuclear bombs. Hydrogen is, in fact, by far the most abundant gas in the Sun, and it is continually being converted into helium with the resultant loss in mass and gain in energy. No elements in the Sun are unknown on earth; and although most of the elements found on the Earth have also been detected in the Sun, by far the largest proportion of the Sun's outer layers consists of the lightest elements, hydrogen and helium.

It is worth recalling here that helium was discovered spectroscopically on the Sun by Lockyer in 1869, twenty-seven years before it was found on the Earth as a very rare gas. Now, of course, it is commonplace and has many practical uses.

We can only see the "outside" of the Sun, the bright surface of which is called the "photosphere". The density increases enormously and rapidly towards the centre, and we therefore see only a short distance below the surface. This is why we see what appears to be a well-defined disk; the disk we measure as being 864,000 miles across. The density increases so rapidly that quite a short distance under this surface the gases are opaque, thus giving the appearance of a solid disk of light. We also know the composition of the outer layers, or what we might call the Sun's "atmosphere", but this will be dealt with later.

The Sun rotates on its axis, but not as a solid body. Its axis is inclined to the plane of the Earth's orbit (the ecliptic) by an angle of $7\frac{1}{4}$ degrees, and it rotates on this axis once in about 25 days. As the Earth is going round the Sun in the same direction, any spot on the Sun, used to time the speed of rotation, will take another two days to "catch us up" and the Sun's period of rotation *as seen from the Earth* is about 27 days. We say "about" since, as indicated above, the Sun does not all rotate at the same speed, as a solid body would. The equator goes round appreciably faster than the higher latitudes, but we must confess that we do not know why this is so.

At about 40 degrees latitude, both north and south, one rotation takes place in about 27½ days while still nearer the Sun's poles, at latitudes of about 80 degrees, the period is about 30 days. The mean periods adopted are 25·38 days for the Sun itself, corresponding to latitudes of about 15 degrees N. and S., or 27·27 days as seen from the Earth.

The Earth is flattened at its poles by its relatively rapid rotation due to the centrifugal force acting on the equatorial region; but either the Sun is not flattened at its poles, owing to its slower rotation, or else the departure from a true sphere is so very slight that it is beyond our detection.

As the axis of the Sun is tilted 7¼° to the perpendicular to the Earth's path, and the Earth's axis is tilted 23½ degrees to this path, the Sun's axis appears to be at different angles, as seen from the Earth, at different times of the year. Each pole can be as far as 26½ degrees first towards the east and as much to the west six months later, while the tilt first backwards from us and then towards us is 7¼ degrees. Fig. 4 shows the appearance of the Sun as we see it with the naked eye, from our northern hemisphere, when it is high in the sky with the north at the top and west to the right. It is of course tilted to the left as it rises and dips to the right as it sets, while southern observers will have to turn the diagram upside down to see the correct orientation. When using optical instruments various orientations are obtained; these are mentioned later.

Surface details belong to the next chapter, but it will

34

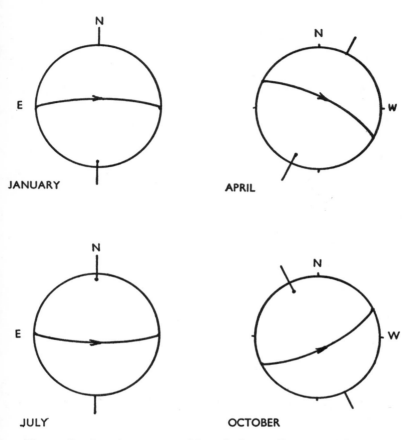

Fig. 4. Sun's axis as seen with naked eye (from our Northern Hemisphere) at beginnings of months shown. In June and December the equator appears straight. Southern Hemisphere observers should view this diagram upside down.

be understood that the apparent paths of any spots on the Sun, across its visible disk, will vary with the time of the year, quite apart from other considerations.

It has been found by spectroscopic means that the Sun has a general magnetic field in the same way as the Earth has one, the Sun's magnetic poles being a few degrees from the axis of rotation. This general magnetic field is not large, but very intense local magnetic areas develop from time to time, particularly in association with the disturbances which manifest themselves on the Sun's surface in the form of sunspots.

As to whether the Sun's total radiation of heat varies, no very definite conclusion can be stated. What is known as the "solar constant" is the amount of heat which would fall perpendicularly per square centimetre per minute on a surface just outside the Earth's atmosphere, and making all allowances for "this and that" only very slight differences have been detected, these apparently being related to sunspot activity, but may be due to other causes also. We only know that any changes in the solar constant must be extremely small and cannot, in themselves, make any appreciable difference to us on Earth.

Above the visible surface of the Sun—the "photosphere" or "sphere of light"—there is a thin layer of gases known as the "chromosphere" or "sphere of colour", so called from its reddish or pink hydrogen with its prominences, and surrounding all is the pearly-white "corona" which is such a beautiful sight during the brief moments of a total eclipse of the Sun and which extends to very great distances. These outer layers of gas can only be explored with special instruments as described in Chapter Seven, while our knowledge of the interior of the Sun can only be theoretical and therefore lies beyond the scope of this book.

Chapter Three

VISIBLE APPEARANCE

THE SUN RISES IN THE EAST, passes south (high in summer, low in winter) and sets in the west like all the natural heavenly bodies, due to the daily rotation of the Earth on its axis from west to east. This must be qualified for southern hemisphere viewers who see the Sun at its highest in the north, but it still rises in the east and sets in the west.

Except when seen through a mist or fog, the human eye cannot stand the brilliance of the Sun, and it is dangerous to look directly at it. To see its disk it is necessary to look through a dark glass filter or an over-exposed piece of photographic film or plate. I have found it convenient to blacken a strip of glass (a 3-inch by 1-inch microscope slide glass is admirable) by holding it over a candle flame to collect the soot. By carefully stroking it across the flame one can make a fair "graduated wedge", light at one end to dark at the other, by trial and error while testing on the Sun; one can wipe off the deposit and try again and again until satisfied. A similar plain glass slip can then be mounted over the treated surface, with a frame of thin card between and lantern slide binding round the edges. A permanent and useful graduated filter is thus obtained, enabling one to see the Sun in comfort at all elevations and seasons; should a naked-eye sunspot happen to be present it will readily be seen, but a spot has to be a very large one to be observed without optical assistance—at least three times the diameter of the Earth and not too near the Sun's limb at the time—and such spots are not very common. Any spot, followed from day to day, will be seen to cross the

37

Sun's disk from left to right (east to west) as it is carried round by the Sun's rotation on its axis.

When we come to observe the Sun through a telescope we see much more detail, and while telescopic methods of observation are dealt with in a later chapter I must mention here that on no account should the Sun be looked at directly through a telescope of any but the smallest size, and even then only with a dark filter on the eyepiece; otherwise irreparable damage will be done to the eye by the intense heat concentrated at the eyepiece.

Many writers have stressed the point to the extent that they refer to *all* telescopes "however small", but I am a bit of a heretic perhaps and, writing as I am for amateur astronomers with common sense, I will confess that I frequently look at the Sun through a small terrestrial telescope of the three-draw type having an object glass of only 1½-inch aperture with a dark glass fitted securely over the eyepiece. I find this telescope most convenient to make a low-power survey of the Sun's face when it is out of range of my observatory due to houses, trees, etc. I can also fit a dark glass filter over one eyepiece of my 8 × 30 binoculars, taking care to fit the filter cap on the left eyepiece which I apply to my right eye as a monocular. The Sun's heat from the right eyepiece then just misses my right ear! This arrangement has proved useful on holiday during sunspot activity when large spots have sometimes appeared.

The inexperienced should disregard the above paragraph on the principle of "safety first", but I do not see why an intelligent amateur astronomer should be deprived of seeing the Sun through a small telescope, or binoculars, having an object glass of 1½-inch or less. Anything larger is admittedly dangerous, as the filter (which should be of dark glass and *not* photographic film in this case) could splinter with the heat, and damage the eye before one could turn aside, however quickly, but with the small apertures mentioned

I have not found the concentrated heat nearly sufficient to cause trouble. The essential thing is to see that the filter cap fits tightly on the eyepiece so that it cannot possibly fall off unexpectedly.

In this chapter we are considering the visible appearance of the Sun only by ordinary methods, seeing by all the visible light and not by selected wave-lengths, which is a subject for the next chapter. We will now consider what we can see with a telescope, bearing in mind that the usual astronomical telescope inverts and reverses the naked-eye view. However, since we must not look direct at the Sun through such a telescope, it should be noted that through a sun diagonal (referred to later), or when the image is projected, it is seen reversed mirror-wise. This question of orientation is discussed more fully later on and I would only make it clear at this stage that my solar photographs in this book are similar to the naked-eye view with N. at the top and E. to the left.

Since the best way to observe the Sun is undoubtedly by projecting the image, Fig. 5 (a) will be a guide to the way this can easily be arranged for a small refractor. Telescopes and observing are dealt with in detail in later chapters, so at the moment we will just consider the principle as illustrated. The cardboard square over the telescope object glass is to provide the necessary shade from direct sunlight, the dotted square at the eyepiece end being an alternative arrangement; the first method has the advantage that the cardboard helps to balance the projection screen mounted at the other end. If necessary a small counterweight can be added to bring the telescope to near balance. The projection screen can be mounted in any convenient way, but the attachment to the telescope should be capable of rotation round the telescope tube, i.e. its axis, for convenience in orientating the image.

An eyepiece is chosen to give a solar image on the screen

of about 4 inches diameter (in the case of small telescopes), but any eyepiece consisting of cemented lenses should be avoided or it may be damaged by the concentrated heat. The Huygenian eyepiece is eminently satisfactory for the purpose. This consists of just two separated plano-convex lenses as in Fig. 5 (*b*).

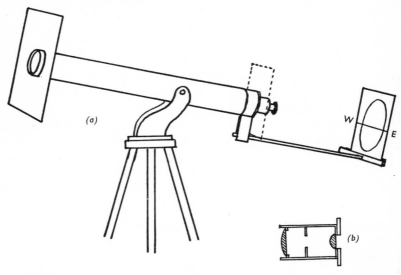

Fig. 5. Projecting the Sun. (*a*) Projecting the Sun with a small telescope; (*b*) Huygenian eyepiece.

The distance of the screen from the eyepiece is best left to trial and error as this, with the power of the eyepiece, will fix the image diameter—for a given object glass focal length. If the image is sharply focused on the otherwise shaded screen a bright image of the Sun's disk will be displayed, showing the main features now to be described. A better arrangement, of course, is some sort of projection box as will be described in Chapter Eight.

LIMB DARKENING. It will be seen at once that the centre of the Sun's disk is brighter than those parts close to its limb, or edge. The light fades quite appreciably the further we look from the centre; this is due to our looking a shorter distance through the outlying gases when observing perpendicular to the Sun's surface at the centre of the disk, whereas when we look towards the limb we are looking diagonally through the outer gases and cannot see so far into the Sun's hotter, denser and brighter layer below. See Fig. 6.

GRANULATION. When the air is steady, giving a well-defined image of the Sun, and using a fairly high magni-

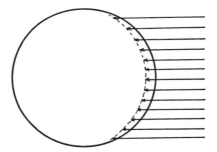

Fig. 6. Seeing further into the hot
Sun nearest the centre of the disk.

fication, the Sun's surface is seen to have a mottled appearance. This is called "solar granulation" and is best seen with a projected image; if one is a little doubtful of its presence it is useful to tap the telescope gently, when the slight vibration of the image seems to draw attention to the minute "rice grains", as they have been called. Although these granulations seem to be very minute bright specks, separated by darker divisions, they are really quite large areas compared with the sizes we are used to on Earth. They vary in size, but average about 500 miles in diameter —which is quite a considerable rice grain!

There is evidently a very turbulent convective layer on the surface of the Sun, and the granulations would seem to be the tops of columns of rising hot (bright) gases. They are in constant movement, changing their appearance in a matter of minutes, and they are more crowded together near sunspot groups.

SUNSPOTS. Spots on the face of the Sun were first seen regularly by Galileo about 1610, in the newly invented telescope, but ancient Chinese records show that occasional naked-eye spots were noted many centuries before this.

Very tiny spots, like minute black dots without any lighter area, or penumbra, are known as pores, and the larger spots develop from these small beginnings. The dark centre, or umbra of a large spot is apparently black only by contrast with the exceedingly brilliant surrounding photosphere. This excess light must be cut down by filters or by projecting the Sun's image, and this correspondingly cuts down the light of the umbra. The temperature of the central umbra is about 4,000 degrees C—only about 2,000 degrees below that of the photosphere—so it is only a little less hot, and brilliant, than the surrounding surface of the Sun. When the opaque disk of the Moon passes across the face of the Sun at the time of a solar eclipse any sunspots that may be visible appear somewhat grey compared with the intense blackness of the Moon.

There are theories as to what sunspots are and how they are formed, but, apart from saying that they seem to be rather like cyclonic storms with expanding gases, resulting in a drop in temperature, I do not want to indulge in theories in this book.

Surrounding the dark umbra of a spot is the lighter penumbra. This more or less follows the shape of the umbra in the case of the fairly regular, or symmetrical, spots, as will be seen from the various sunspot photographs in this book, but it can surround whole groups of spots of the irregular kind and cover a vast area—as shown in Plates XX–XXIV, for instance. The larger spot groups frequently take the form of long streams extending to great distances.

The largest sunspot group recorded so far appeared in 1947 and covered an area of over 7,000 million square miles; about a hundred Earths could be fitted into this area.

The penumbra of a spot seems to be the solar granules turned over towards the centre and elongated. Sometimes

(a)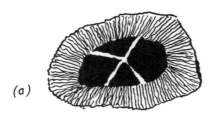

(b)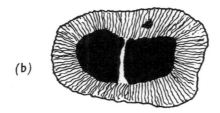

Fig. 7. Sunspot with photospheric "bridges", showing change in 24 hours. (a) 1949 May 14; (b) 1949 May 15.

the photosphere or bright surface of the Sun, itself forms bright bridges across the spots, as will be seen in Plates XVII and XIX, and I have sketched in Fig. 7 a very striking bridge formation as I saw it in 1949, showing the complete change in 24 hours as recorded at the time. Occasionally a bright "island" appears in the middle of the dark umbra as seen in Plate XVI (top).

As the spots are carried across the Sun's disk by its rotation, those in different latitudes do not move at the same rate since the higher latitudes of the Sun's surface rotate at a slower speed than its equator. This means that for groups of spots that last long enough to cover more than one rotation there will be a change in their relative longitudes

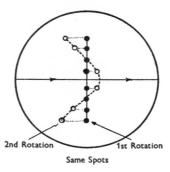

Fig. 8. The Sun's different rotation speeds at different latitudes.

when they come into view a second time. This is illustrated in Fig. 8.

The outstanding and puzzling feature of sunspots is the very remarkable periodicity in their abundance—they come and go in a regular way every eleven years (approximately). Individual spots form, grow, decay and disappear in a matter of days; or, in the case of long-lived spots, they may last for many months, disappearing round the west limb of the Sun and reappearing round the east limb some two weeks later, the Sun's period of rotation being 27 days as seen from the Earth, as mentioned earlier. But the *frequency* of the spots as a whole varies from periods when the disk is entirely free from spots for days or weeks on end (sunspot minimum) to periods in between when the Sun is always spotted with a number of groups scattered across its face, many being quite large (sunspot maximum). This "sunspot cycle", from minimum to maximum and back to minimum, always takes approximately the same time. The average is actually 11·2 years, but some variation takes place from cycle to cycle. The rise to maximum is usually more rapid —3½ to 4 years—than the drop to minimum—7 to 7½ years.

A great sunspot maximum occurred towards the end of

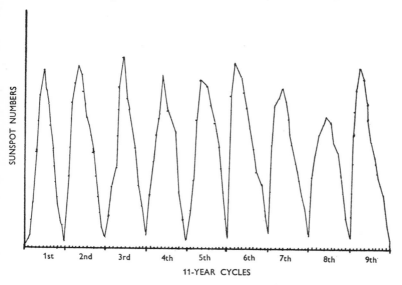

Fig. 9. Typical graph of sunspot activity.

1957, and the International Geophysical Year (I.G.Y.) was fixed to coincide so that solar effects on the Earth, and in our atmosphere, could be studied when they were having the most effect. The last maximum occurred in 1969.

Plate IV is a particularly fine photograph of the whole disk, taken when the Sun was at its maximum activity. It shows the sun "thoroughly spotted".

The 11-year cycle is all very mysterious and has given rise to a number of theories, but it cannot yet be said that the true cause is known. It is all these unknowns that make solar studies so fascinating. It is interesting to recall that it was a German *amateur* astronomer, Schwabe of Dessau, who discovered the 11-year cycle as a result of regularly studying the Sun's face, whenever it was visible, over a period of 25 years. Before that time (1826–51) no such periodicity was even suspected. The frequency and areas of sunspots are now regularly recorded, and a typical graph of frequency over the years is shown in Fig. 9.

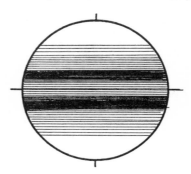

Fig. 10. Sunspot zones.

Another peculiarity of the sunspot cycle is the change of latitude of spots as the cycle progresses. Spots are never found near the Sun's poles and not very frequently on its equator. With a new cycle commencing, i.e. rising from minimum, the first sparse spots appear in the higher latitudes, both in the northern and in the southern hemispheres, but rarely beyond 40 degrees and never, one might say, higher than 45 degrees. The average is about 30 degrees north and south. As the cycle proceeds towards maximum, the spots get more frequent and they appear in areas somewhat nearer the equator until, at sunspot maximum, the average latitude N. and S. is about 15 degrees. Then as the cycle develops towards sunspot minimum, the spots appear less frequently, but they continue their approach towards the equator, fading out at minimum at about 7 degrees latitude.

At the same time as the old cycle spots are disappearing at low latitudes the new cycle spots in the high latitudes begin to make their appearance, so there is a certain overlap of the cycles. Fig. 10 shows the bands of sunspot activity

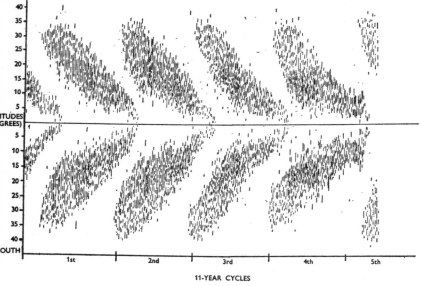

Fig. 11. Representation of "Butterfly" Diagram.

and how they are infrequent at the high latitudes and at the equator, and most prolific at about 15 degrees N. and S.

The changes in sunspot latitudes over the cycles and the overlap of the cycles are well illustrated in the so-called "butterfly diagram" obtained by plotting spot positions against years as depicted in Fig. 11. This is called the butterfly diagram for obvious reasons, and was first plotted by

Maunder, of Greenwich Observatory, in 1904. It has been brought up to date since.

The "blackness" of the umbra of a spot has been mentioned, but occasionally differences in darkness have been reported and some observers have recorded reddish patches. These should be watched for.

As to the profiles of individual spots, they are generally considered to be saucer-shaped cavities in the Sun's surface,

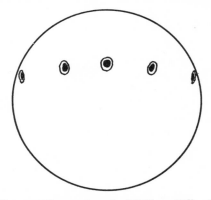

Fig. 12. Illustrating the "Wilson Effect".

but ideas in the past have been rather misleading. Dr. Wilson, of Glasgow, made some observations in 1769 of a large sunspot which was then approaching the Sun's west limb. He followed its changing appearance and recorded the shape and proportions of its penumbra, which was at first equally broad on both sides of the umbra. As the spot was carried nearer to the west limb by the Sun's rotation he noticed that the penumbra nearer the Sun's centre became narrower, and when the spot was very near indeed to the limb the contracted penumbra had vanished from sight while the width of the penumbra nearest the limb remained practically unaltered. Dr. Wilson concluded that the spot was a vast and deep cavity in the surface

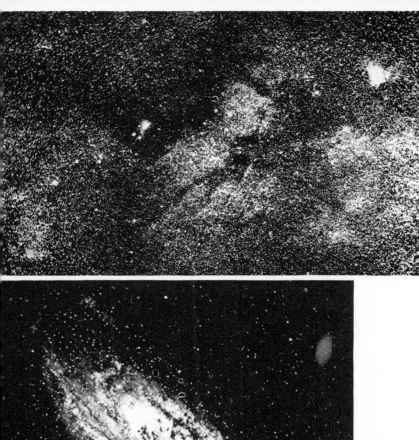

Plate I.   (*a*) A part of the Milky Way.
           (*b*) Galaxy in Andromeda (M31).

Plate II. (a) Galaxy in Ursa Major (M81).
(b) Galaxy in Coma Berenices (N.G.C. 4565).

Plate III. Total eclipse, 1961 February 15, photographed by H. C. Hunt at Pisa.

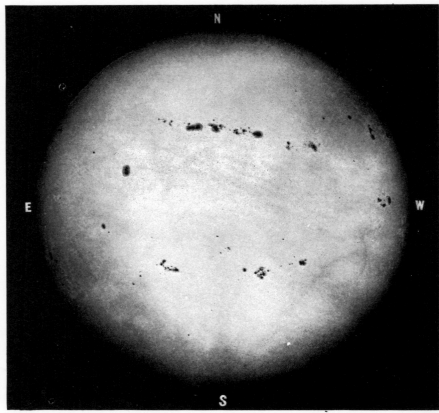

Plate IV.   The face of the Sun, 1957 December 25,
photographed by Hans Arber (Manila) when the Sun's
activity was at a maximum.

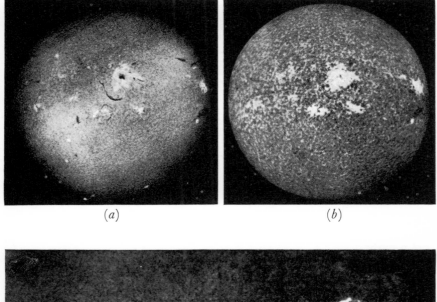

(a) (b)

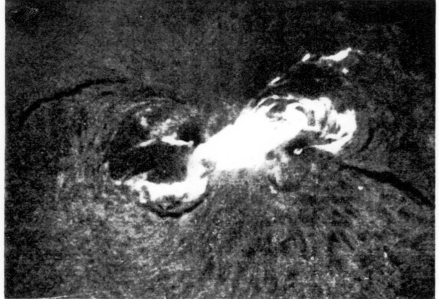

Plate V. (a) Spectroheliogram in hydrogen light.
  (b) Spectroheliogram in calcium light.
  (c) A solar flare.

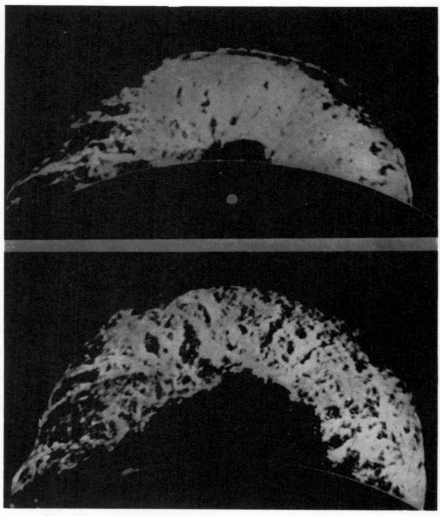

Plate VI. A gigantic arched prominence, 1946 June 4.
(*a*) The small white disk shows size of Earth.
(*b*) Photographed half an hour after (*a*).

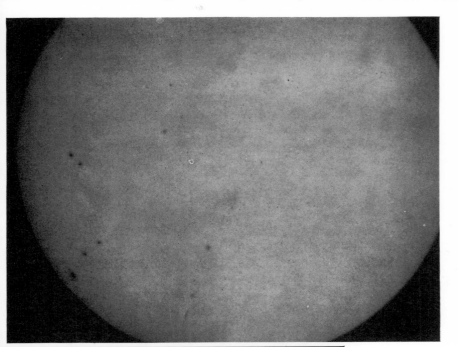

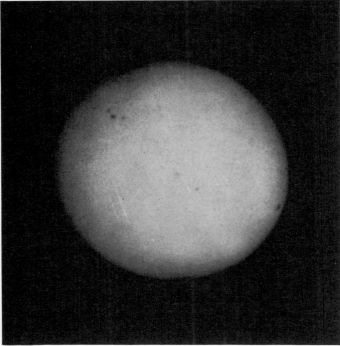

Plate VII.   (*a*) Photograph of Sun through diagonal.
(*b*) Photograph of projected image.

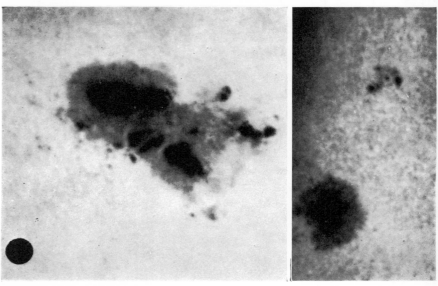

Plate VIII.

(*a*) H. N. D. Wright's camera mounted on 3-inch telescope.

(*b*) Large sunspot (1960 November 13—5·6 inch disk enlarged to 32 inch. Black circle shows size of Earth for comparison.

(*c*) Spot and solar granulation (1960 October 15)—5·6 inch disk enlarged to 35 inch.

of the Sun with the umbra at the bottom and the penumbra its sloping sides; saucer-shaped, in fact.

His surmise was confirmed when the same spot appeared again at the east limb some two weeks later and he noted that the position was reversed. The narrow penumbra was still on the side nearest the Sun's centre while that near the limb was wide, as before, although the spot was now on the opposite side of the disk. This change in the perspective of a spot came to be known as the "Wilson Effect" and is illustrated in Fig. 12.

Since that time it was generally accepted that *all* spots were hollows of considerable depth, the average quoted by various writers being anything from 500 miles to 6,000 miles or more. An investigation into the "Wilson Effect" is being carried out at the present time by the Solar Section of the British Astronomical Association (B.A.A.) under the guidance of its Director, H. Hill, and members of the Section take an important part by supplying suitable sunspot photographs taken for this specific purpose and is carrying out the necessary calculations in collaboration with the Director. Only "regular", or symmetrical, spots are suitable for this study and examples of both a "hollow" and a "mound" are shown in Plate XIII.

All that can be said at the moment is that sunspots in general are found to be relatively shallow depressions of a few hundred miles, and occasionally one would appear to be a mound rather than a depression. When the investigation is completed and a really large number of spots have been measured and their depths, or heights, calculated, it is hoped that the statistics will indicate once and for all the general profiles of spots and show that the earlier writers had much exaggerated ideas of spot depths—ideas which continued for so long without any real check having been attempted. This investigation is a good example of co-operative work that can still usefully be carried out by amateurs.

Incidentally, some of the older observers have maintained that they have seen hollow spots as notches in the Sun's limb as they were appearing or disappearing round the limb. I have never seen such a notch, and there is some doubt as to whether it would be optically possible to see one as it would normally be hidden by the higher boundary of the penumbra nearest the centre of the Sun. Moreover, even if half of the umbra were still visible on the observer's side of the limb, it would hardly be distinguishable from the dark sky background. Photographs that I have seen are not conclusive, to my mind. However, with a very elongated (E. to W.) umbra it might just be possible to see a small "bite" in the limb and it would be well worth while looking for. Let us hope that someone succeeds in clearly photographing such an effect; I have certainly been searching for the opportunity for a long time, but without any luck so far.

A new sunspot usually develops from what is seen first as a tiny pore. As the days go by it enlarges and a definite penumbra forms. Further spots frequently break out in the vicinity and the whole can become a large sunspot group. On the other hand, some pores come and go without developing at all and daily observations show new pores coming and going. The larger spots may last for weeks or months, appearing regularly at the E. limb but changing shapes and sizes; one group is on record as lasting for about 180 days. These changes are fascinating to observe and record. Some spots form on the hidden side of the Sun, of course, and are seen for the first time as they come round the E. limb, so one never knows what will be seen new on the Sun's disk when turning a telescope on to it for the first time each day.

Sunspots also have a way of appearing in pairs, and there is one obvious connection between the members of such a pair—their magnetic polarity. This subject should not really be mentioned under the heading of "visible appearance", but it seems the best place to draw attention to another important property of sunspots. It was mentioned earlier that very strong local magnetic fields occur with active sunspots and the spectroscope enables us to measure their intensity and ascertain their magnetic polarity by observing the splitting of the lines in the solar spectrum, referred to in the next chapter (Zeeman Effect), but this is not a technique for the amateur.

Members of a pair are of opposite polarity and the two

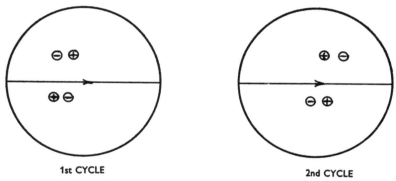

1st CYCLE   2nd CYCLE

Fig. 13. Magnetic polarity of sunspots.

polarities of pairs in the Sun's northern hemisphere are opposite to those of the pairs present at the same time in the southern hemisphere. In each case the spot (of the pair) nearest the west, i.e. in the direction of the Sun's rotation, is called the "leader" and the other one of the pair the "follower". If in the Sun's northern hemisphere the leader of a pair is of positive, or north, magnetic polarity then its follower is found to be of negative, or south, polarity, while for another pair, visible at the same time but in the Sun's southern hemisphere, the leader is negative and its follower positive. The very strange thing is that this state of affairs is completely reversed at the beginning of each new 11-year sunspot cycle. Taking the above example, during the next cycle it would be found that the northern hemisphere leader would be negative and its follower positive, while the southern hemisphere leader would be positive and its follower negative. This is made clearer by Fig. 13.

Single spots have the polarity of "leader" spots in the same hemisphere, and in the case of large complicated groups of spots the magnetic polarities become complicated too, the group containing a mixture of polarities.

These magnetic changes are indeed strange and, again, are not fully understood, although all theories of sunspots, and of solar activity generally, must take them into account. The outstanding feature is that the Sun really has a 22-year cycle of activity and not one of 11 years. Less evident cycles, some less and some greater than 11 years and superimposed on the main cycle, are suspected and a constant watch is kept for definite evidence.

It seems clear that the differential rotation periods of the different parts of the Sun must cause enormous internal stresses and electric currents which give rise to the intense magnetic fields found in the disturbed areas on the surface. These in turn cause turbulent storms which appear to slow down the convection currents and, with expansion, cause somewhat cooler patches with resulting darkening, i.e. sunspots. However, I must not start theorizing myself, as this book is intended to be practical rather than theoretical.

FACULÆ. When examining the Sun through a telescope we see besides the limb darkening, the granulations and any sunspots that may be present, irregular and extensive bright patches known as faculæ. These appear to be "hot" active areas of hydrogen, just above the surface of the Sun. These light areas are not generally visible against the bright photosphere near the centre of the disk, but they show up very plainly near the Sun's limb, where the

darkening enables them to be seen by contrast. See Plates
XVI (bottom), XVII and XIX (bottom).

Faculæ are not confined to the sunspot latitudes, but when
they are seen in these tracts and are very bright they are
usually a precursor of an outbreak of sunspot activity in
their midst. In fact active sunspots are usually surrounded by
irregular patches of faculæ although, as mentioned above, the
latter are not normally visible except when such spots are
near the Sun's limb. Faculæ are not frequent towards the
Sun's poles, but when present they are usually in the form
of small but bright patches.

In the next chapter we shall pass from the "visible" to
the "invisible"—invisible in an ordinary telescope, that is
to say.

Chapter Four

RADIATIONS FROM THE SUN

THE OBVIOUS RADIATIONS FROM the Sun are its light
and heat. These are electromagnetic waves travelling in
space at the speed of 186,000 miles a second, and they reach
us from the Sun, 93,000,000 miles away, in 8½ minutes.
These electromagnetic waves are of various wave-lengths,
and therefore of various frequencies, and only a very small
range is visible to the eye as light. Other radiations—of
shorter wave-length than visible light and of longer wave-
lengths than visible light—are detectable by instruments,
but not by the eye. One can think of the analogy of sound,
bearing in mind that sound waves cannot travel in a vacuum
(space), but only through air. The human ear can only
detect a limited range, but there are shorter sound waves
(higher notes) and longer sound waves (lower notes) that
cannot be heard, but can be detected by other means. In
fact sounds beyond the range of the human ear can nevertheless be heard by some animals and insects.

White light (sunlight) consists of a band of wave-lengths
from the shorter waves giving the sensation of violet light
through all the colours of the rainbow—indigo, blue, green,
yellow and orange—to the long visible waves giving the
sensation of dark red light. In fact raindrops have the property of splitting up the sunlight falling on them under
certain conditions to form the glorious rainbow, displaying
to us the full range of the coloured "spectrum". It is this
property of white light—consisting of a mixture of all colours
—that colours the objects all around us. A red flower has
the property of absorbing all the colours in sunlight except
the red waves, which are rejected and reflected to our eyes,

giving us the sensation of seeing a red flower. In the same
way a blue flower, or any other blue object, reflects only the
blue waves to our eyes. This is why monochromatic light,
or light of just one colour, can only show in their true colour

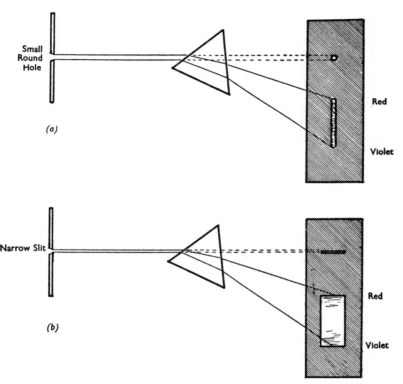

Fig. 14. Formation of spectra. (a) Spectrum formed with circular
hole; (b) spectrum formed with thin rectangular slit.

those objects which have the property of reflecting that
colour. Pure yellow light, for instance, will show up yellow
objects brilliantly but not other colours. One only has to
look at one's companions in the light from sodium or mercury street lamps. Most of the colours of the spectrum are

55

missing in these cases: complexions look muddy, and bright clothing drab—the colours are just not in the light from the lamps and cannot therefore be reflected to our eyes from the objects which we know are actually coloured.

Isaac Newton, in 1666, was the first to split up white light into its component colours; he did this by allowing a beam of sunlight to enter a darkened room through a small circular hole in a shutter. He placed a glass prism (a triangular block of glass) in the path of the beam of light and found that as it was "refracted" by the prism it was also bent or "deviated". But he also found that the shorter (violet) rays were deviated more than the longer (red) rays—with intermediate colours between—and this fanning out of the rays, or "dispersion", could be displayed on a screen as depicted in Fig. 14 (a).

The range of colours was called a "spectrum", and with his circular hole to admit the sunlight Newton was forming an "impure" spectrum, as he was obtaining a series of images of the circular hole, each of a different colour and overlapping, but giving the effect of a "continuous spectrum"—continuous from the violet to the red without any demarcation. It was much later that Wollaston, another Englishman, separated the wave-lengths more precisely by substituting a narrow straight slit for the circular hole in the shutter, thus obtaining a "pure spectrum". It was rather like trying to investigate the range of notes on a piano (each with its own wave-length) by first putting a board on the keys to depress them all and then finding the separate notes by individual fingers. The difference was not obvious because the continuous spectrum looked the same, but when light of restricted wave-lengths was examined in the same way the narrow slit produced individual lines of light—coloured images of the slit—instead of a confused overlapping. This will be seen more clearly from Fig. 14 (b).

When Wollaston examined sunlight with his narrow slit

and prism in 1802, long after Newton had made his discovery, he noticed a number of fine dark lines crossing his spectrum, at right-angles to its length, but he did not understand their significance and did not pursue the matter. It was Fraunhofer, a German, who in 1814 thoroughly investigated this phenomenon, mapped some hundreds of dark lines and gave his name to them—Fraunhofer lines—by which they have been known since. He realized that the band of colour of the continuous spectrum was made up of side-by-side (but continuous) images of the slit, and that the dark lines were gaps in the spectrum or what we might call missing images of the slit, or missing wave-lengths. The positions of the lines showed which wave-lengths were absent from the Sun's light and he found that the same lines appeared in the light from the sky, the clouds, the Moon and planets as they were all reflecting sunlight. Later on the light of the stars was examined and it was found that dark lines were present but their positions differed for the different stars and from those in sunlight. Certain lines in the solar spectrum were more pronounced when the Sun was low in the sky, shining through a greater length of our atmosphere, and it was subsequently found that these were due to absorption in the atmosphere and would have to be taken into account when studying the spectra of heavenly bodies. It was later still, in 1859, that Kirchhoff and Bunsen gave the full explanation of the dark Fraunhofer lines; that an element will absorb just those very wave-lengths of light that it, itself, emits. The dark lines are known as "absorption lines".

When chemical elements are heated in a laboratory to incandescence they give out light of restricted wave-lengths and when examined with the slit and prism show individual bright lines known as "emission lines"—each having its own set of lines; rather like individual "signature tunes", to use the sound analogy again. We will return to this in a moment.

A spectroscope is a refined instrument for producing and examining spectra, but fundamentally it consists of a slit and prism. More is said about this instrument in Chapter Seven, but it must be mentioned here that another way to disperse light into a spectrum is by substituting a "diffraction grating" for the prism. This consists of a number of fine lines ruled very close together on glass (some 14,000 or more to the inch), which, with its property of light interference, gives much the same effect as a prism, though with some differences. Incidentally, interference colours are seen in thin films such as oil on water, a soap bubble and, to the annoyance of colour photography enthusiasts, in transparencies mounted between glass where pictures are sometimes spoiled by the appearance of "Newton's Rings" in the most unwanted places. This can now be overcome by using specially treated cover glasses.

It is easier to obtain greater dispersion with a grating than with a prism, but with some loss of light, so that it is usual to use a prism, or prisms, on the fainter objects such as stars and to use a grating on the Sun where light is plentiful. Moreover, a diffraction grating gives a well-spaced-out spectrum with the line positions in proportion to their wavelengths, making it more convenient to measure the wavelengths of specific lines. By comparison the prism spectrum is drawn out at the violet end and compressed at the red end.

The two types of continuous spectra are shown, one below the other, in the coloured Frontispiece (*a*) and (*b*); (*c*) is a representation of the solar spectrum showing just a few of the chief dark absorption lines. Actually the solar spectrum has thousands of such dark lines crossing it from end to end, indicating the elements in the Sun's "atmosphere" which are absorbing certain specific wave-lengths from the bright (continuous spectrum) background of the Sun's hotter interior—for, as remarked earlier, an element absorbs just those very wave-lengths of light that it, itself, emits.

If sodium, for instance, is heated to incandescence it gives out a yellow light, and if this is examined with a spectroscope it is found that the light is really restricted to a very narrow band in the yellow part of the spectrum, no other colour being visible. With sufficient dispersion it will be seen that the yellow band really consists of two thin yellow lines very close together. If one now produces a bright continuous spectrum with a strong (hot) white light and then places cool sodium vapour between the source of light and the slit of a spectroscope, it is found that two narrow dark lines (absorption lines) appear across the continuous spectrum in *exactly* the same position (the same wave-length) as the bright emission lines given by the incandescent sodium. These sodium absorption lines have been sketched in on the Frontispiece (*c*) and (*d*).

In the same way that the analytical chemist can tell what elements occur in an unknown substance by heating it and examining its bright emission spectrum, each element having its own "signature tune" or lines of particular wave-lengths, so can the astronomer examine the Sun and see from the positions of the dark absorption lines what elements are present in its outer layers. In this way we learn the constituents of the Sun's "atmosphere" dealt with in our next chapter.

Returning to our subject of radiations from the Sun, we have now looked at the visible range of electromagnetic waves, that is to say the visible spectrum, but beyond the short-wave violet end there are even shorter waves invisible to the eye as light, but which, nevertheless, can affect a photographic plate. Photographs can be taken by ultra-violet light, although to the eye there is complete darkness. From the photographic point of view ultra-violet rays are very useful to the astronomer but, fortunately for us, we receive them on the Earth, from the Sun, in a very mild form or they would be really damaging to our bodies. We all know how we can get excessively sunburnt when exposed too long to these injurious rays—perhaps at the seaside or up high mountains. The Sun radiates very powerful ultra-violet rays, X-rays and cosmic rays, but they are largely intercepted by layers in our upper atmosphere. They

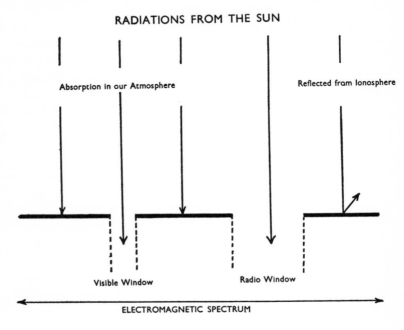

Fig. 15. The visual "window" for optical astronomy and the radio "window" for radio astronomy.

are a hazard that has always to be faced by space travellers going beyond the protective layer by which we are surrounded.

Beyond the ultra-violet in the spectrum are the gamma rays, while, at the other end of the spectrum beyond the long waves of visible red light, are the heat rays which can be felt and measured although not seen, while beyond those are the short-wave radio and then the long-wave radio radiations.

In the total length of the known solar spectrum of electromagnetic waves the visible portion is only a very narrow band of wave-lengths—a narrow "window" through which the optical astronomer can survey the universe. There is another "window", but of somewhat wider range, in the region of radio radiations and this enables the radio astronomer to study the heavens with his own special electronic equipment. It is through these two windows of the spectrum that all our astronomical knowledge is obtained—the optical window, by our sight, from the time of man's first appearance on the Earth, and the radio window, by instrument, only from very recent times. See Fig. 15.

The Sun is a very important and interesting subject for study by the radio astronomer, and for more information on this topic I cannot do better than refer my readers to more technical and specialized books.

As will have been understood, the *optical* astronomer observes in only a very narrow band of wave-lengths, between about 4,000 Ångström units at the violet limit of the spectrum to about 7,000 Ångström units at the red end. Ångström units (Å) are the measure of wave-length, 1Å being equal to 1/100,000,000 of a centimetre, so they are very short units indeed; but wave-lengths shorter still are those of the ultra-violet, X-rays and gamma rays. Beyond the red the *radio* astronomer deals with the longer waves, measured in centimetres and metres.

Fraunhofer designated letters of the alphabet to the chief absorption lines he saw in the solar spectrum, and these are indicated by the white letters in the Frontispiece just above (*c*). The white lettering at the bottom indicates the corresponding elements responsible for those lines and the following are the respective wave-lengths in Ångström units:

Fraunhofer line	Wave-length (Å)	Element
C	6563	Hydrogen
D₁	5896	Sodium
D₂	5890	,,
F	4861	Hydrogen
H	3968	Calcium
K	3934	,,

When the Sun is particularly active the radiations we receive are enhanced. The electromagnetic waves take $8\frac{1}{2}$ minutes to reach us, and these include visible light, but there are sometimes sudden outbursts of radiation which are not visible to us in normal "integrated" light—that is to say light of all wave-lengths as integrated into the solar continuous spectrum—but are visible only in certain specific wave-lengths which require special instruments to detect them, such as the spectroscope, the spectrohelioscope or the coronagraph as referred to in Chapter Seven. With these instruments we can study outbursts of solar energy which reach us in $8\frac{1}{2}$ minutes from the time they leave the Sun. The radio astronomer records a burst of radio "noise" on his instruments at the same time as the optical astronomer observes them, since radio waves travel at the same speed as light, but the disturbed Sun can also eject electrified particles and these travel at a slower speed, taking 20 to 40 hours to reach us.

One of the chief sources of these outbursts of solar energy, or solar storms, are the "flares" with their intense ultra-violet and cosmic-ray radiation. A solar flare can appear quite suddenly on the Sun's surface and may fade away in an hour or so, the greatest intensity lasting for perhaps only a few minutes. There have been one or two historic cases of flares being of such intensity that they were actually visible in an ordinary telescope in integrated light, the most famous being the observation made of a particularly brilliant flare in 1859 by Carrington. Normally flares are only visible in the monochromatic light of hydrogen, with the special instruments mentioned, and one would be very fortunate indeed to repeat Carrington's observation; still, one always hopes when a particularly active group of sunspots appears on the face of the Sun. Flares are usually associated with spots,

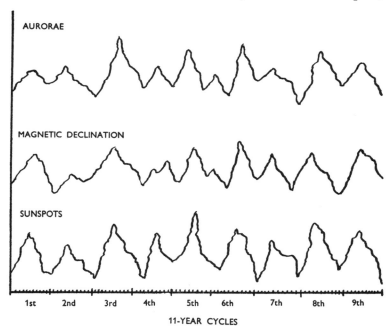

AURORAE

MAGNETIC DECLINATION

SUNSPOTS

| 1st | 2nd | 3rd | 4th | 5th | 6th | 7th | 8th | 9th |

11-YEAR CYCLES

Fig. 16. Sunspot numbers, Earth's magnetic declination and number of auroræ observed.

appearing unheralded in the midst of very active groups, and this is where they should be looked for—particularly if the observer is the fortunate possessor of a spectroscope.

At the same time as a flare is seen and radio noise is recorded there occurs a fade-out of our short-wave radio transmissions. These fade-outs are a great inconvenience, to say the least, to our commercial transmissions, and sometimes give interference problems to our television engineers. Sometimes a TV announcer will apologize for a poor picture due to "causes beyond our control", occasionally putting the blame where it belongs—on sunspots, but really on flares.

The electrified particles which arrive a day or so later curve towards the Earth's magnetic poles as they approach us and give rise to a glow in our upper atmosphere giving those beautiful coloured lights in the sky, the Aurora Borealis in the Arctic regions and the Aurora Australis in the Antarctic. When there has been a really big outburst of electrified particles from the Sun these auroral displays have been truly magnificent spectacles in our higher latitudes and have been seen almost to the equator. They are certainly most impressive when visible on dark, cloudless and moonless nights away from city lights. At the same time our instruments register intense magnetic storms which, again, can upset our communications. Fig. 16 illustrates the relation between sunspot activity, magnetic variations and displays of the auroræ, during a number of solar cycles. I should perhaps explain that Figs. 9, 11 and 16 are purposely sketched as only general representations of the graphs to show the principle as clearly as possible. Plots of actual solar activity in specific years are available in well-known books on the Sun.

Mention was made in Chapter Three of the International Geophysical Year (I.G.Y.) fixed for 1957-8 to coincide with a period of maximum solar activity, and scientists are still evaluating the mass of information obtained. There was also an International Year of the Quiet Sun (I.Q.S.Y.) during the period 1964-5 so that a real comparison could be made of the effects of the Sun on the Earth and its surroundings during maximum solar activity and those during minimum activity.

The effects I have mentioned as being registered by us on the Earth naturally make us wonder how our daily lives may be affected in other ways by the Sun's activity. We have seen already that our life on Earth is entirely dependent on the Sun and that the changes in the "solar constant" are so slight as to be inappreciable in its effects on us. The rings seen in cut sections of tree trunks certainly indicate by their variations that there must be some slight changes affecting their growth, corresponding to the 11-year solar cycle—perhaps due to slight changes in the humidity of our atmosphere—but we really cannot blame sunspots for local effects such as our changeable weather, earthquakes, volcanic eruptions, floods and other earthly catastrophes with which we are inflicted from time to time. Neither can we put responsibility on the Sun for our bad holiday weather! While the effects of the Sun are considerable from the scientific point of view, the "man in the street" is only likely to complain that radio transmissions are sometimes affected, or even out of action for a short time, and his TV picture may get temporarily distorted. On the other hand he may see some fine displays of the Aurora. He cannot have everything!

Chapter Five

SURROUNDING "ATMOSPHERE"

NAKED-EYE VIEWS OF TOTAL eclipses of the Sun are magnificent spectacles, but they are very rarely seen from any one country, and to reach the path of totality, that is to be within the narrow band of the Moon's shadow as it tracks across the Earth, usually involves travel to more or less distant parts. We were fortunate indeed in having a total eclipse in February 1961 which was visible across southern France, central Italy, Yugoslavia and southern Russia, the shadow very conveniently crossing some major astronomical observatories on its way. Those of us in Europe who could not travel to the actual path of totality were treated to television pictures of the phenomenon for the first time in history. Plate III, already referred to, is a fine photograph taken by H. C. Hunt from Pisa while Plate XII (bottom) is a photograph of the same eclipse as seen in France, but taken of my television screen with an ordinary hand camera—what one might call "science from an easy chair"! The former is of course much the better picture, but the latter is included for comparison and as a matter of interest; I think it is a credit to the efforts of Eurovision in this new field.

I have a happy recollection of seeing the total solar eclipse of 1954 from a site in Sweden. The outstanding feature seen with the naked eye on these occasions is undoubtedly the beautiful white halo—the Corona—surrounding the black disk of the eclipsed Sun. A telescopic view will show a thin ring of pinkish light surrounding the Sun's limb, which is the "chromosphere" or layer of "atmosphere" immediately above the visible surface of the Sun. Rising here and there from the chromosphere will usually be seen pinkish protuberances or "prominences". The coloured layer of chromosphere and its flame-like prominences emit the pinkish light of incandescent hydrogen, as they mainly consist of this gas.

Before solar studies with the spectroscope were possible, a total eclipse of the Sun attracted astronomers to distant parts of the world as it was necessary to be as near as possible to the centre of the Moon's shadow to obtain the longest time available to observe what is a very brief phenomenon. Totality lasts only a matter of minutes, seven minutes at the most, but usually less than this. Moreover, the path of totality has a way of crossing most inaccessible places, so that it was not uncommon for arduous expeditions to be undertaken only to find that during the brief moments of totality the phenomenon the observers had gone so far to see was obliterated by cloud.

It is now possible for those parts of the Sun's surroundings which were only observed at the moments of total eclipse to be seen and photographed on any clear day. The coming of the spectroscope gave added purpose to eclipse expeditions and, although the dark Fraunhofer absorption lines in the solar spectrum had been studied for some time, it was the India eclipse of 1868 that gave astronomers a leap forward by showing that not only could the bright hydrogen emission lines of the chromosphere be seen during totality, but with the more powerful spectroscope then available it was possible to see these emission lines when the eclipse was over, by placing the telescopic image of the Sun's limb tangentially on the slit of the spectroscope.

Lockyer in this country had reasoned that if only the brilliant sky illumination immediately surrounding the Sun's limb could have its intensity sufficiently reduced, while the brilliance of the emission lines in the spectrum was maintained, it should be possible to see the lines against a darker background, although in full sunlight. The idea was to have a more powerful spectroscope by increasing the amount of dispersion, thus extending the continuous spectrum to a really great length compared with that obtained with the earlier, relatively simple, instruments. The continuous spectrum being made up of the sum total of all the wave-lengths of visible light could, by greater dispersion, be more drawn out and its brilliance be correspondingly reduced. On the other hand, the thin spectral lines were of certain specific wave-lengths only and drawing out the spectrum would not reduce their intensity but only take them further apart from each other. As the dispersion was increased the background grew darker with the emission lines remaining bright, thus enabling them to be visible even in full sunshine.

A French astronomer, Janssen, had the same idea quite independently when observing the India eclipse. He was so struck by the brilliance of the hydrogen emission lines seen when all but the chromosphere was hidden by the Moon that he endeavoured to see the lines again when full sunlight had returned, and he succeeded. He was delighted to find that he could also see them again on the following day. Although Janssen was the first actually to see the emission lines without an eclipse, Lockyer's independent work to the same end was recognized and a joint medal for them was struck by the French government to commemorate the occasion. Normally we only see the Sun's total visible light, or what is called "integrated light", but it was now possible to isolate the light of a particular wave-length and observe the Sun, or selected parts of it, in "monochromatic" light as emitted by specific elements in the Sun's atmosphere.

The Sun's chromosphere and its prominences could now be seen on any clear day, but at first only in the form of a narrow bright hydrogen line in the spectrum. The depth of the chromosphere and the shapes and sizes of prominences could only be ascertained by scanning with the slit bit by bit until a full picture could be built up. A little later on, however, it was shown that with still greater dispersion the background light could be so much reduced that the spectroscope slit could be opened wider until a whole prominence, if not too large, could be seen, its shape drawn and its size measured.

It must be remembered that all the lines of the spectrum, whether dark or bright, are images of the slit. If the slit is curved, all the lines are curved, and if the slit takes the shape of a prominence the lines take the same shape. When the telescopic image of a hydrogen prominence is focused on the slit, the bright hydrogen emission lines take its shape and, with a sufficiently dim background and the slit widened, the actual pinkish bright prominence is seen on a broad dark background. This assumes that the Hα emission line of hydrogen, in the red end of the spectrum, at wave-length 6563 Å, is employed as is usual. A blue image of the prominence can be seen in the Hβ line, but this image is not usually so clear and sharp as that in the red line.

The Frontispiece (d) is a representation of the solar spectrum when the spectroscope slit is placed on the image of the Sun's chromosphere—showing the bright hydrogen Hα and Hβ lines. It also shows a bright helium line, not seen as an absorption line.

The Frontispiece (e) gives a general impression of what is seen when the slit is widened so that the shape of the prominence is made visible.

An important property of the spectroscope is its ability to indicate, and measure, speeds of light sources towards or away from us, that is to say the velocities of approach or of recession, on the well-known Doppler principle. Taking the analogy of sound again, when any object emitting a definite note is approaching us the pitch of the note is constant as long as the speed of approach is constant, but when receding from us the same note being emitted falls a tone or so to our ears, the amount of the drop depending on the speed the source is travelling. It is usual to cite the example of an express train with its locomotive whistling as it roars through a station. To the people on the station platform the whistle is constant in pitch as the train approaches, but the pitch falls as it passes, and then remains constant, at the lower pitch, as it recedes. Unfortunately, a train never seems to whistle for me while I am waiting on a station, but I do remember the sense of relief experienced when the drone of the engines of an enemy bomber, during World War Two, dropped a pitch or two and I felt that the plane was no longer approaching but receding.

The explanation is, of course, that as the source of the sound approaches us the sound waves are compressed or shortened, beating on our ears more frequently, and thus we hear a higher note than would be heard if the source were stationary. In recession the waves are drawn out or lengthened, beating on our ears less frequently, resulting in our hearing a lower note. The pitch is really a measure of the wave-lengths, or of the frequency, of the sound striking our ear-drums. High notes are of shorter wave-lengths, and of higher frequency, while low notes are of longer wave-lengths and of lower frequency.

In the same way light, which travels through space (which sound does not), in reaching our eyes through the spectro-scope registers the wave-lengths of the source by the position of the spectral lines. If the source of light is approaching us there is a slight shift of the lines towards the violet end (short waves) of the spectrum, while if the source is receding the shift is towards the red end (long waves). Approach or recession as measured by the change in wave-length of the lines does not indicate whether the source, or we, or both are moving—it is the *relative* movement that is seen and measured.

If the image of the east limb of the Sun is directed on to

the spectroscope slit and then on to the west limb the lines in the two spectra do not coincide exactly. The lines from the east limb (coming towards us as the Sun rotates on its axis) are very slightly displaced towards the violet, while the lines from the west limb (going away from us) are very slightly shifted towards the red. The spectra can be put side by side, the difference in the wave-lengths measured and the speed of the Sun's rotation ascertained. This method is a check on the rotation periods observed by the movements of sunspots across the Sun's disk.

The Doppler principle is used, too, to measure the speeds of the hot gases of the Sun's chromosphere and of the prominences in the line of sight. Movements at right-angles to our vision can be recorded by just watching or photographing the changes, but the spectroscope is the only instrument by which we can measure velocities of heavenly bodies towards or away from us. This is most important when studying movements in the Sun's atmosphere.

The spectroheliograph and the spectrohelioscope are dealt with more fully in Chapter Seven, but the principle will be mentioned here.

Although the absorption lines in the solar spectrum appear dark, by contrast, they nevertheless admit sufficient light to leave a record on a photographic plate, in the light of the particular wave-length chosen, when the rest of the spectrum is completely blocked out. If an image of the Sun is allowed to traverse the spectroscope slit and the hydrogen Fraunhofer line is isolated by a second slit, then a photographic plate behind this second slit will receive only hydrogen illumination from the Sun. By arranging to move the plate along at exactly the same speed as the Sun's image traverses the first slit, a photographic picture of the Sun in hydrogen light will be built up on the plate. Everything appearing in the photograph will be hydrogen and nothing but hydrogen. If one of the violet calcium lines is chosen for the same purpose then a picture of the Sun showing calcium, and calcium only, will be obtained. The same results can be obtained if the Sun's image is kept stationary, by the telescope clock drive, and the photographic plate fixed. The two slits then move together, the first traversing the Sun's image while the second passes across the plate. This is the principle of the spectroheliograph invented by Hale in America in 1892 and, independently, by Deslandres in France.

Plate V shows photographs (spectroheliograms) of the Sun taken by this special technique (*a*) being taken in hydrogen light and (*b*) in calcium light.

Some thirty years later Hale followed up his invention with the spectrohelioscope in which both the spectroscope slit (on the Sun's image) and the second slit (isolating the chosen line) are made to vibrate in unison. Thus, while the first slit is rapidly scanning the Sun the second slit—also vibrating and now viewed with a low-power eyepiece—allows, by persistence of vision, one actually to see the Sun in, say, hydrogen light with its hydrogen prominences and flares (if any).

REVERSING LAYER. When the Sun is eclipsed and we see the last rays from the limb as a thin crescent of light as the Moon encroaches on the Sun's disk, these rays must come from the lowest layer of the Sun's atmosphere, and spectroscopic observers see a brief flash of the bright emission lines, each in the form of a crescent following the pattern of the bright image. What were dark lines from the photosphere absorption are reversed when only the chromosphere is visible, and they flash out brilliantly for a few moments each in its own colour. The same "flash spectrum" is seen, of course, when the first rays from the returning sunlight appear at the end of totality. The thin layer of gas immediately above the Sun's photosphere responsible for this change from dark absorption lines to bright emission lines was called the "reversing layer", for obvious reasons, but this term is not generally used now as it is realized that a separate layer is not involved, but simply a part of the chromosphere—the lowest part.

The temperature of this thin layer is lower than that of the photosphere, but as we rise in the chromosphere the temperature increases very rapidly.

CHROMOSPHERE. This layer is next to the photosphere, or visible surface of the Sun, and is a mass of turbulent gas. It is largely an envelope of hydrogen, but with calcium and helium too. It is difficult to say how far the chromosphere extends when the experts do not agree, which again illustrates how much there is still to learn about the Sun. One authority says the chromosphere reaches to a height of 6,000 miles, another to 9,000 miles and a third refers to the upper parts as being some 30,000 miles above the photosphere. The difficulty is that the gaseous layers do not have clearly defined limits and, in any case, the upper layers of the chromosphere merge with prominences which can vary in height enormously.

FLOCCULI. The hydrogen, as we have seen, gives rise to the very prominent line in the red end of the solar spectrum and the calcium is responsible for the two strong lines in the violet. In the hydrogen picture in Plate V (a) are seen white cloud-like patches and similar white clouds are seen even more pronounced in the calcium picture in (b). They are always present in photographs of the Sun taken through the spectroheliograph in the monochromatic light of calcium, and are known as flocculi. They appear to be "hot clouds" in the chromosphere, above the photosphere with its sunspots and faculæ.

PROMINENCES. These are the really big and fascinating hydrogen "flames" that are a joy to the amateur spectroscopist. The layer of the chromosphere immediately surrounding the Sun's limb is seen to be turbulent, with little "flames" called spicules coming and going all the time. But quite frequently really large prominences can be observed stretching away from the Sun for thousands of miles. An average height might be between, say, 15,000 and 30,000 miles, but extreme cases might be 500,000 miles or more— that is more than a solar radius. Such have been recorded.

A really wonderful arched prominence of colossal size was photographed in 1946 and this is shown in Plate VI (a) and (b), showing the change which took place in half an hour.

Prominences fall into two main classes, although a number of subdivisions are recognized. First we have the "quiescent" ones, and these, while not of great height, persist for a long time with little change, perhaps for weeks. Then there are the "eruptive" ones, and these are prominences which give the striking displays of luminous gases rising to great heights at great speeds. Not only do they rise but they also descend from great distances sometimes, as if a downpour were coming from the corona above. Moreover, many of these active prominences which shoot upwards in curves return to the Sun's surface back along the same curves, and not straight down, indicating that they must be guided back by magnetic or electromagnetic forces and not just by gravity.

These interesting features can be watched in the open slit of a spectroscope as represented in the Frontispiece (e) and it will be appreciated, from what has been said, that they are seen as bright pinkish "flames" in the monochromatic light of the hydrogen of which they are composed. Records are kept, in the same way that sunspots are regularly recorded, and a complete picture over the years is obtained—again by co-operative effort.

Do not expect to see the rapid movements of the eruptive prominences while you "look and stare" into the spectroscope. The Sun is so far away that a movement of 8,000 miles in an hour, for example, would mean that the top of the prominence would rise only by a distance equal to the diameter of the Earth in one hour. But the Earth alongside the Sun would be a mere dot, less than one-hundredth of the Sun's diameter, and such a movement would hardly be perceptible within the hour. And yet, who would say that 8,000 miles an hour is not a rapid speed? Actually the gigantic prominence shown in Plate VI rose at a speed of about 400,000 miles an hour (more than 100 miles a second) and in such an exceptional case the motion would be visible over a relatively short period of time. When the second photograph was taken it had already reached a height of about 120,000 miles and it continued still higher until it reached outward to about a solar diameter before breaking up. This was one of the greatest solar explosions ever recorded.

Several of the professional solar observatories now keep a regular cinématograph film watch on solar activity with automatic cameras, and by taking time-lapse motion pictures at 30-second intervals and subsequently viewing them at the normal projection rate—hence speeded up—striking and graphical records of solar changes are obtained. Most of us have seen some of these exciting ciné pictures either at scientific meetings or in TV science and school-programmes.

Prominences seen at the Sun's limb are bright flame-like gases, quite brilliant against the dark sky as seen by their restricted wave-lengths of hydrogen, but when they are observed on the extremely bright face of the Sun they appear as dark filaments, as will be noticed in Plate V (a). In numbers they follow, in general, the sunspot cycle, but to a less pronounced degree, and although they are most prolific in the sunspot bands of activity they do appear at high latitudes, even occasionally at the poles.

FLARES. These were mentioned in the previous chapter as being the source of the strongest radiations received from the Sun and those which affect us most. They come un-heralded, but we know they are most likely to appear in the midst of very active sunspot areas so that if such a sunspot group comes round the east limb, by the Sun's rotation, our radio engineers must be on the look-out for possible troubles arising in their transmissions.

Flares come quickly and go more slowly, but in any case they do not last for very long periods, as do sunspots. They are another subject for fruitful study with the spectroscope and similar instruments and time-lapse motion pictures automatically record their birth, rise and fall. Plate V (c) is a photograph of a major flare taken in hydrogen light by the same technique as used for Plate V (a).

A number of observatories in various parts of the world keep a constant "flare patrol" and provide useful informa-tion not only to astronomers but also to the communication engineers who are concerned with radio interference pro-blems.

The amateur with a solar spectroscope is indeed fortunate if he "catches" a flare in his instrument, but automatic photography ensures that any outbreaks are not missed, while observers with spectrohelioscopes keep an eye on the active Sun, too, and flares readily show themselves in such instruments. All the same, it is worth sweeping particularly active areas with the fixed slit of the ordinary solar spectro-scope in the hope of striking a flare.

Flares are classified in order of importance, a combination of size and brilliance, as Class 1, Class 2 and Class 3, with the addition of a + or − to indicate more, or less, than average. The most important are designated 3 +. Although their origin is not yet known they would appear to be some form of very intense luminous electric discharge, and they are certainly the source of very powerful ultra-violet radia-tion. Individually, their lives are relatively brief, but they can give rise to eruptive prominences and yet leave their background of sunspots, faculæ and granulation much as they were, when the "excitement" is over.

CORONA. This, the most striking part of a total eclipse as seen with the naked eye, is still very much of a mystery. The earlier observers of eclipses thought it was merely an optical effect due to our atmosphere, while some even thought it belonged to the Moon, but it was soon realized by eclipse expeditions that it is a very real appendage of the Sun. Moreover, its shape varies from eclipse to eclipse, changing with the sunspot cycle. At sunspot minimum the corona has

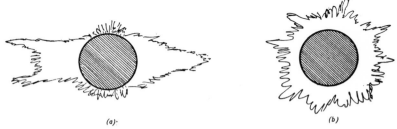

(a).   (b)

Fig. 17. Typical shapes of the visible corona. (a) At sunspot minimum; (b) at sunspot maximum.

long streamers extending from the Sun's equatorial regions to vast distances, with short brush-like streaks adjacent to the poles. At sunspot maximum the corona is much more regular round the whole periphery of the Sun. A general representation of these two extremes is given in Fig. 17 and it will be understood that between sunspot minima and maxima the corona takes intermediate shapes. The sketch is only to illustrate the shapes in a general way as they vary to quite an extent from time to time.

It is also impossible to give a true representation of the relatively faint illumination which stretches out further and further into space, getting fainter and fainter, without any definite ending. Even a photograph cannot compete with the human eye, which can take in a wide range of illumination,

77

as a plate correctly exposed to show detail in the inner corona which is relatively bright cannot show the outer extensions. Similarly a long exposure to show the faint outermost parts grossly over-exposes the inner corona; the total light of the whole corona is only about that of the full Moon. However, Hunt's photograph in Plate III is an excellent picture of what was seen at the 1961 eclipse.

It is easy to appreciate that the corona is very tenuous indeed, and what is most remarkable is that its temperature, as ascertained by several techniques, is found to be extremely high—in the order of 1 million degrees. It is difficult to understand how there can be such a high temperature so far from the source of heat, the Sun itself; but nevertheless we must accept that such is the case. We are told that in such a tenuous "atmosphere", bordering on a complete vacuum, the coronal electrons are moving so rapidly that the random Doppler shifts of the absorption lines of its spectrum are blurred out and are not seen, thus indicating the high temperature mentioned. The corona shows certain characteristic bright emission lines suggesting very high ionization energy, again pointing to the same high temperature. The extent of the tenuous corona also supports this conclusion. Moreover, the radio astronomers agree, too, from their investigations in their own range of wave-lengths. It depends, of course, on what you mean by "temperature" —that which we measure with a thermometer under our normal conditions or by the speed of electrons—but I must leave this problem to the experts, not being a physicist myself. In any case the corona is not an object for observation by the ordinary amateur optical solar observer, except at the time of total eclipse when it will certainly be studied if only as a magnificent sight.

It is now possible for the professional astronomer with special equipment to photograph the corona without waiting for a total eclipse, the special equipment being the coronagraph invented by Lyot, mentioned in Chapter Seven, but even this is limited to the brighter parts of the inner corona.

I would only add that the radio astronomers, studying the Sun through their own "window", can trace the extremely tenuous outer corona to much greater distances than can the optical astronomers, either visually or photographically. In fact they tell us it may well stretch further into space than the Earth's orbit so that from that point of view we, on the Earth, live in the borders of the Sun's "atmosphere" as well as in our own.

Chapter Six

THE AMATEUR'S TELESCOPE

I AM SURE THAT READERS WILL be familiar with the
telescope, but, nevertheless, I will risk criticism by at least
outlining the principles of the refracting and of the reflecting
telescopes, both types being in the hands of "ordinary"
amateur astronomers.

The principle of telescopes in general is very simple. If a
lens, of a foot or so focal length, say, is allowed to catch the
sunlight, a small bright image of the Sun can be formed on
a piece of paper held at a distance of the focal length behind
the lens. It is a bright and hot image, as every schoolboy
knows who has had a "burning glass" directed to his neck by
a colleague! To avoid heat troubles try doing the same with
the Moon, when a small bright image will be projected on
to the paper. Then stand behind the paper (which should
be thin to allow the image to be seen from the back) and
focus a small, shorter focus lens, such as a hand-magnifier,
on to this image, and an enlarged image of the Moon will
be seen. The paper can be removed without destroying the
telescopic view, as the image formed by the first lens is there
all the time "in the air". In fact the Moon will be brighter
and clearer, and we now have a simple refracting telescope.

We can form a similar image of the Sun or Moon by
means of a concave mirror in place of the first lens, but in
this case the paper must be a small strip held in front so as
not to hide the mirror. Here the difficulty is to avoid obstruc-
ting the mirror with one's head when looking at the image
through the small magnifier, and the mirror will have to
be tilted. Here we have the simple reflecting telescope, and
the makers of such instruments overcome the difficulty

mentioned by means of a small flat mirror, at 45 degrees, projecting the light to one side, both mirrors being surface-coated to avoid double reflections. This is the principle of the "Newtonian" reflecting telescope invented by Isaac Newton about 1668. These two types—the refracting and the Newtonian reflecting—are the ones in most general use by amateur astronomers and, while there is an extremely wide variation for all sorts of special purposes, we will content ourselves with what most amateurs have or are likely to acquire. Some beginners might, in fact, ask why they should bother with anything other than the simple refractor.

The refractor was invented by Hans Lippershey, a Dutch optician, in 1608—and was shortly afterwards used to view the heavens constructively by the famous Italian Galileo— had a simple convex lens as its object glass or "objective" (the lens nearest the object being viewed) and this will produce a fringe of colours around an image which will also be ill-defined. This will be clear from what has been said about the formation of the spectrum in Chapter Four. In fact a bi-convex lens (both sides curved outwards) is not unlike two acute-angled prisms placed base to base and it will be appreciated that the blue rays are then refracted most and come to a focus nearer the object glass, as shown in Fig. 18 (a). The red rays form their image further away. The various colours in the white light received from the object glass are focused at different planes, and it is impossible to obtain a really clear picture.

This disadvantage of a refractor was overcome by the invention of the "achromatic" objective by the Englishman Dollond in 1757. Based on the principle that some types of glass deviate light-rays more, but disperse them less, than other types (other densities) it was found possible to correct the colour dispersion by making a compound lens of two glasses of different densities and yet not entirely cancel out the deviation. This is shown for simple base to base prisms in Fig. 18 (b), which is also the principle of the direct vision prism spectroscope referred to in the next chapter, but in this latter case we are doing the opposite—correcting the deviation while maintaining dispersion.

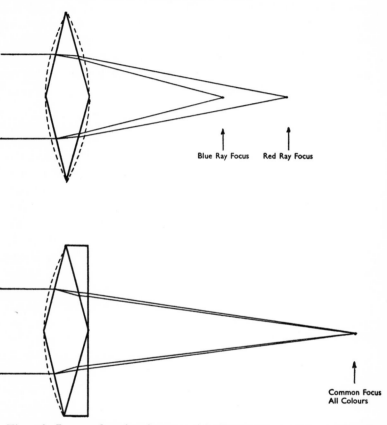

Fig. 18. Lenses forming images. (a) Convex lens—blue and red foci; (b) achromatic lens—common focus.

The achromatic objective gives practically a colour-free and sharp image, but for a good lens other corrections have to be made too. It will be appreciated that I have

82

over-simplified its construction, as the grinding, figuring and polishing of telescope lenses is a very skilled "art" and one not to be tackled by the ordinary amateur.

When we consider the reflecting telescope, however, we are in the province of the amateur telescope constructor, as

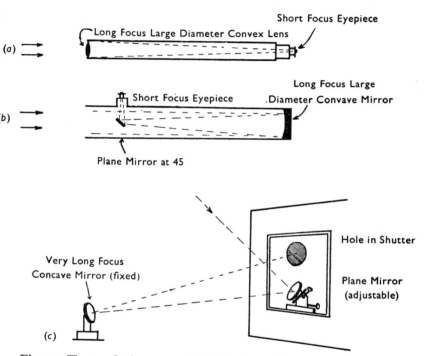

Fig. 19. Types of telescopes. (a) Refracting telescope; (b) reflecting telescope; (c) long focus sun telescope.

only one slightly concave mirror surface of glass has to be figured and the glass itself need not be optically perfect—as an objective has to be to *transmit* its light perfectly, whereas now we are dealing with light that is *reflected* from the surface only. There are many books giving full instructions for grinding, figuring, polishing and metal-coating telescope mirrors and for constructing reflecting telescopes, and amateurs are fully capable of turning out quite large diameters —up to, say, 12 inches—with good performance on the heavenly bodies. I even constructed a 6-inch myself!

There are two main advantages of the reflector over the refractor. Firstly, one obtains a really perfect colour rendering with all the colours of the spectrum focused in one plane; the light-rays are reflected from the surface of the mirror and do not suffer any dispersion by passing through the glass. This is a particularly useful property when studying delicate tints such as are seen on the surfaces of the planets. Secondly, the reflector is much cheaper than the refractor, size for size, when bought commercially and, as I have said, it can be home-made by any enthusiastic "do-it-yourself" amateur.

TELESCOPES. Fig. 19 (*a*) is a sketch of a refractor showing the path of rays from the object to where they converge to form the image of that object at the eyepiece where it is magnified. Fig. 19 (*b*) shows the Newtonian reflecting telescope with the rays converging from the concave mirror to the eyepiece via the small flat mirror (which could be a totally reflecting glass prism if preferred).

In studying the night sky, astronomers desire a maximum amount of light as well as magnification and a clearly defined image. The amount of light gathered depends on the diameter of the objective or of the mirror, whereas the size of the image formed depends on its focal length. Large diameter, bright image. Small diameter, faint image. Long focal length, large magnification, but faint image. Short focal length, small magnification, but bright image. But, very important, the ability to "resolve" fine detail, or see as separate objects very close planetary markings or double stars, for instance, depends on the diameter. So it is a question of a compromise between brightness, magnification and resolution. In a camera one can alter the relationship between diameter of lens and focal length by varying the "stop" or F-number. Refractors usually work at about F/15, that is to say the focal length is fifteen times the diameter of the objective, while reflectors are usually much shorter compared with their diameters—more like F/8.

Astronomical telescopes invert and reverse the image compared with the naked-eye view, but this is not a real disadvantage. To give an erect image, as in a terrestrial telescope, it is necessary to insert additional lenses between the objective and the eyepiece and these would only cut down the light and possibly impair the definition somewhat. Astronomers are so used to seeing through their telescopes, and drawing what they see, with the image inverted that it looks quite odd to see a picture of, for instance, the Moon with the prominent crater Tycho near the bottom. Once you are used to working with an astronomical telescope and its movements there is no real drawback to this reversal. Having said all that, may I remind readers, as an aside, that my sunspot photographs in this book are all the "right way up", with north at the top! You see, the camera again reverses.

In observing the Sun we have an abundance of light, and we do not require such a large diameter objective or mirror as for seeing the fainter objects of the night sky. We want it sufficiently large to give good resolution, but we can have a longer focal length, with a correspondingly larger and fainter primary image, and the loss of light can be accepted and can even be deemed an advantage. Observatories do not need 200-inch diameter telescopes to see the Sun. Instruments specially devoted to solar work are of moderate diameter, but very long and hence inconvenient to move, so they are fixed with the light of the Sun directed to them by supplementary flat mirrors. Some are fixed horizontally, but the more usual arrangement is to have a vertical telescope in a "solar tower" on the top platform of which are the object glass and the mirrors for collecting the light and guiding it down the telescope tube. These are covered by the usual dome with opening shutters. At the foot of the fixed telescope, perhaps some distance underground, the primary image of the Sun—of large diameter—is formed where it can be examined by spectroscope or photographed in a well-equipped and temperature-controlled laboratory.

Fig. 19 (c) sketches an arrangement for projecting the Sun as suggested in *Amateur Telescope Making*, that excellent *Scientific American* publication which is such a useful guide to the amateur astronomer who can make things. All that is required are two 6-inch mirrors, one flat and the other concave. The concave mirror can be of very long focal length indeed—50, 75 or even 100 feet—requiring very little hollowing out and producing a very large image of the Sun, even larger than the diameter of the mirrors; about 11 inches diameter at 100 feet. If projected into a darkened room through a hole in a shutter the Sun can be studied by a roomful of people. I mention this "Sun telescope" as it seems that we have here a simple arrangement that should be relatively easy to construct by any amateur with plenty of space, but I confess I have not tried it, nor have I heard of anyone in this country who has. Perhaps there are snags.

STANDS. Turning to telescope stands, or mountings, I cannot emphasize enough how important is a steady mounting for any satisfactory work to be done with any telescope, and the larger the instrument the more important this is. So often one sees a fine telescope which cannot be used to full advantage owing to "shake". Even the amateur who spends so much time and patience grinding, figuring and polishing a mirror, and constructing the telescope itself, is sometimes inclined to skimp the mounting, while some of the commercial stands for the smaller refractors are not too good in this respect. It is better to err on the heavy side and make the stand more robust than perhaps is really necessary.

Many amateurs' telescopes are mounted on a simple altazimuth stand as illustrated in Fig. 20 (*a*) for a refractor and
in (*b*) for a reflector, the base being on a tripod for the
smaller refractors or on an iron pillar for the larger instruments. This stand has the advantages of cheapness and sim-
plicity, but whenever possible an equatorial stand should be
made or acquired. This is illustrated in Fig. 20 (*c*) for a refractor and in (*d*) for a reflector, although some reflectors
have a mounting similar to (*c*). It must be borne in mind
that the eyepiece of a refractor is low down and the telescope
should be mounted fairly high, whereas in the Newtonian
reflector the eyepiece is high and the mounting should be
kept low.

The owner of an altazimuth stand has to give his telescope
two movements to follow any heavenly object as it passes
across the sky due to the Earth's daily rotation. When the
object is rising the telescope must be turned horizontally
round the azimuth axis and, at the same time, be elevated
by the altitude axis. As the object passes south and goes
towards its setting point the telescope must be slowly dipped
while continuing the azimuth motion. In the southern hemisphere, for "south" read "north".

With high magnifications, which also magnify the apparent motion of the heavenly bodies across the field of view,
this double motion can be tedious, and the only satisfactory
way to study detail is to set the telescope in advance of the
object and let the latter drift across the field of the stationary
telescope, mounted, we hope, on a really steady stand or it
will suffer from wind vibration.

It is quite impossible to impart these two motions by
mechanical means, and when a clock-drive is required an
equatorial stand is, of course, a necessity; clock-drive mean-
ing any form of mechanical or electrical drive.

It will be seen from Fig. 20 (*c*) and (*d*) that the two axes
of an equatorial stand are at right-angles to each other,

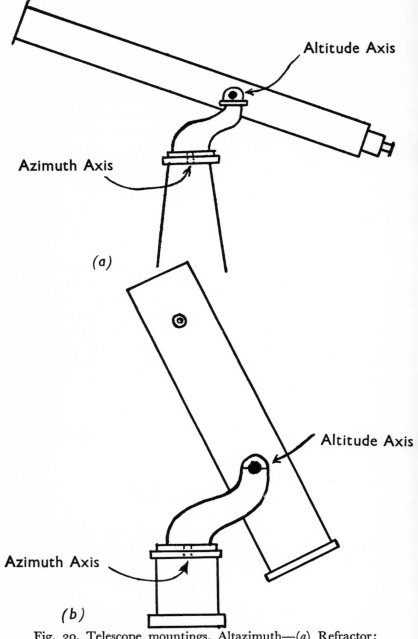

Fig. 20. Telescope mountings. Altazimuth—(a) Refractor;
(b) Reflector.

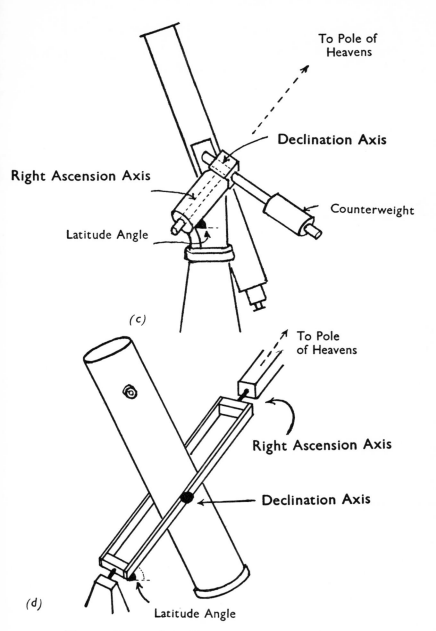

Fig. 20. Equatorial—(c) Refractor; (d) Reflector.

and the right ascension axis is inclined towards the pole of the heavens, or of the celestial sphere, by an angle to the horizontal equal to the degrees of latitude of the observing site. In other words this axis lies *exactly* parallel to the earth's axis, and it will be understood that, whereas an astronomer in our northern hemisphere has to point his right ascension axis to the north pole of the heavens, the southern hemisphere astronomer points his axis to the south pole of the heavens. On the equator the axis will be horizontal.

With this type of stand, if the telescope is set on to an object as it rises, with the declination (or height) axis then clamped fast, the telescope can follow the object as it rises, passes across the sky and sets, by just the one motion only, turning on the R.A. axis. It will be clear that a clock-drive can now be coupled to this axis, enabling any object to be held in the field of view for hours at a time. It is true that no automatic drive is perfect and a touch on the controls is necessary from time to time (a particularly close watch must be kept during long photographic exposures on faint objects), but such drives are certainly a great convenience.

Other adjuncts, which many amateurs maintain are unnecessary, are the two "setting circles", one on each of the two axes of an equatorial mounting, which enable a telescope to be directed to any object whose position is given in tables or in star maps, or to an object invisible to the naked eye such as a planet in daytime.

Circles are really a necessity for the last mentioned purpose, and they are usually fitted to commercially produced equatorial stands; certainly for the larger telescopes. Amateurs can readily fit home-made circles to their home-made telescope stands, of sufficient accuracy to enable objects to be found with a low power, large field, eyepiece. Circles, if one happens to have them, are most useful in accurately setting up an equatorial stand—and absolute accuracy is essential—and mine proved very helpful when erecting my own telescope and, subsequently, for picking up objects, particularly Venus in full daylight.

Perhaps it would be of some use to some amateurs if I set down the procedure for the accurate erection of an equatorially mounted telescope fitted with circles:

1. *Collimation Error.* Select a star near the equator and the meridian. Turn telescope to E. of pillar (stand). Note the time and R.A. reading. Turn telescope to W. of pillar. If difference of the two readings of R.A. less 12 hours differs from the time interval between the two observations the telescope is not truly at right-angles to the declination axis. If the difference is greater than the time interval then, with telescope E., the object glass must be moved away from the pillar, i.e. to E. Insert thin foil under telescope cradle as necessary.

2. *Declination Axis.* Turn to star on meridian, telescope E. of pillar. Read declination circle. Turn telescope to W. of pillar. Read declination circle again. Take average of readings and, setting telescope E., reset verniers of circle to mean reading. Repeat with higher star near meridian and again on several stars until satisfied. Turn to a high star near meridian, set circle to read *apparent* position of star and set telescope axis for elevation.

3. *Azimuth.* Choose two stars of nearly equal declination, one E. and the other W., both fairly high. Set declination circle for W. star. Turn R.A. axis until star is in centre of field (or on web or crosswire). If unable to do this, turn in azimuth until star is in centre. Then turn to E. star. Repeat, but note how much telescope is turned in azimuth. Move in azimuth half back and azimuth should be true. Repeat several times, noting declination circle readings for each star after making the adjustment. The error due to refraction of the atmosphere should be the same for both stars if they are at approximately equal altitudes and declinations.

4. *Hour Circle.* Take the time when a star near the meridian crosses the web and move R.A. circle verniers if necessary to correct.

I think it was Dr. Steavenson at a B.A.A. meeting who caused much laughter when he said he knew someone who used circles to find the Sun. It was said as a joke, of course, and he was not referring to me at the time, but let me confess that I have done just that! Perhaps I have been going to pick up Venus in the daytime, or search for Mercury, and as a check I have first set my circles for the Sun—before even opening the observatory shutter—and then on opening up have had the satisfaction of seeing the Sun in the centre of the field. Of course there are occasions when it does not work out this way as it is very easy to make mistakes such as in calculating the correct local sidereal time (from the B.A.A. *Handbook* or *Whitaker's Almanack*) during Summer Time! It is a useful exercise, anyway, and harmless fun, so why should not one practise on the Sun?

For solar observations quite a small telescope can be used. A 2-inch object glass refractor with suitable eyepiece will project an image of the Sun's disk 3 or 4 inches in diameter, quite bright enough to show any of the larger sunspots that may be present. A 3-inch telescope is better, and a 4-inch is perhaps best of all as it gives ample illumination for projecting quite a large disk, with good resolution, and yet does not cause the excessive heat concentration that a big instrument would.

Larger telescopes usually have their objectives stopped down, and owners of reflecting telescopes also stop down their larger mirrors to about 4 inches. This diameter seems to be that most frequently adopted by amateur solar observers. Owners of reflecting telescopes are sometimes advised to have an uncoated mirror for solar work so that only a small part of the light and heat is reflected from the concave mirror surface to the eyepiece, the bulk of the heat being transmitted straight through the mirror and out of the back of the telescope, but I have yet to meet an amateur astronomer who does not ever want to observe the night sky, or who wants to have two mirrors, one for day and one for night. No, I think the ordinary observer will use his telescope to the best advantage on the night sky, and stop it down to 4 inches, if it is larger than this, for observing the Sun.

The general opinion seems to be that refractors are, on the whole, more efficient on the Sun than reflectors, largely due to heating troubles and resulting air currents in the latter, and a 4-inch equatorially mounted (clock-driven) refractor is probably ideal for the purpose; but perhaps I am biased, being the possessor of just such an instrument. That is not to say that good solar work is not done with reflectors and smaller refractors. After all, it is the man behind the telescope who is really responsible for the results, and a good observer will do more useful work with an "ordinary" telescope than a not-so-good observer will do with a bigger and better instrument.

I must mention that the late F. Addey, former Director of the Solar Observing Section of the B.A.A., regularly observed the Sun with his home-made 4-inch reflecting telescope, plotting sunspot positions whenever the Sun appeared. He ground, figured and polished his 4-inch diameter Pyrex glass mirror and constructed an equatorial stand which, to the casual onlooker, looked very "rough and ready", to say the least, but in the hands of Addey it did good work. This should be sufficient encouragement to amateurs who think they must have large telescopes to do useful observing and hence do not attempt anything worth while with what they consider to be their inadequate resources.

Another regular solar observer operating with a reflecting telescope is H. Joy, of Reading, who uses a 10-inch reflector, but with the mirror stopped down to 4 inches during the summer months. His method of projection is referred to in Chapter Eight.

D. P. Bayley, of Warrington, has a 5¾-inch equatorial refractor and contributes regular drawings, and his projection method is also referred to in Chapter Eight. In addition Bayley makes spectroscopic observations.

M. Atchison, of London, N.W.2., observed with the 6-inch Cooke equatorial refractor of the Hampstead Scientific Society and had the use of a solar spectroscope.

H. N. D. Wright, of London, S.W.17, has a modest 3-inch equatorial refractor with which he takes very successful solar photographs—see Chapter Nine and Plate VIII. His work should encourage others with similar simple equipment.

J. B. Orr, of Auckland, N.Z., has a fine observatory containing a 12-inch Maksutov reflecting telescope on the same mounting as a 4-inch refractor with which he regularly takes excellent solar photographs. See also Chapter Nine.

Hans Arber, of Manila, is fortunate in having an exceptionally well-equipped observatory in the Philippines, and he is able to contribute observations when many other observing sites are beset with clouds. He has a 6-inch refractor with a 4-inch attached, equatorially mounted, and he regularly records sunspots, taking photographs of interesting spots and groups. Again, see Chapter Nine and Plate IV.

Frank W. Hyde used to have a very fine 9-inch refractor with Coudé mounting, at his radio astronomy observatory at Clacton, although such a telescope is unlikely to be in the hands of other amateurs. Being of special construction, I did not feel it warranted mention in the brief account of telescope types earlier in this chapter. It is an equatorial, with clock-drive, and the telescope tube itself forms the R.A. axis pointing upwards to the north celestial pole and terminating with the eyepiece in a fixed position in a wooden observatory. The eyepiece simply rotates in its fixed position as the telescope moves in R.A., thus enabling the observer to sit in comfort, rather like looking down into a microscope on a table. The telescope tube has a hinged elbow joint in the middle and the objective half can be moved up and down in declination by controls inside the observatory. At the elbow a flat mirror automatically directs the light from the objective to the eyepiece whatever the declination.

The arrangement is something like Fig. 20 (d) if one imagines the lower half of that telescope tube to point upwards to form the upper part of the R.A. axis in that figure while the hinged joint (with flat mirror) is where the declination axis is shown. The top of the telescope tube contains the objective, while the bottom half—if one imagines it bent up from the declination axis—contains the eyepiece which now points up the R.A. axis and remains in a fixed position in the observatory, merely rotating slowly as objects are followed by the telescope. It must have been a very comfortable observatory in which to work.

On the inside of the sloping roof, opposite to the eyepiece, Hyde had fixed a white screen on which a large image of the Sun can be projected and this enables him to compare his solar radio observations with what is seen visually on the Sun's face at the time. After seeing his equipment I "borrowed" this idea and can now throw a large projection of the Sun on my observatory roof. See Plate X (b).

EYEPIECES. When we consider telescope eyepieces, or oculars, there is certainly a large variety of types to choose from—single lens, Huygenian, Ramsden, Kellner, Orthoscopic, Tolles and so on—and I suppose they each have their advantages for specific purposes but, having tried a number, I am satisfied that for general solar work the ordinary Huygenian is as good as any and better than most. Some solar observers may not agree, but that is my story and I am sticking to it!

The essential point to bear in mind is that cemented lenses, as some eyepiece components are, should never be used on the Sun, since the concentrated heat is liable to damage them, as previously emphasised.

The Huygenian is a simple eyepiece consisting of two plano-convex lenses each with its flat side towards the eye and having a diaphragm, or stop, between them as shown in Fig. 5 (*b*). The field lens, further from the eye, is the larger to collect maximum light from the object glass or mirror, while the eye lens is smaller and provides the main part of the magnification. The focal plane of the combination is between the two lenses and near the diaphragm. This eyepiece cannot be used as a "positive" lens to focus on to crosswires.

Various telescope magnifications are obtainable by choosing eyepieces of different focal lengths, bearing in mind that the magnification of a telescope as a whole is ascertained by dividing the focal length of the object glass, or mirror, by the focal length of the eyepiece (in the same units, of course). But therein lies the rub. How does one measure the focal length of an eyepiece which is a combination of two lenses with the focal plane "somewhere" between them? I have been amazed at the inaccurate labelling of astronomical eyepieces sold by optical dealers. If the manufacturers would accurately determine their own focal lengths and engrave them on their eyepieces, a lot of trouble and disappointment would be avoided. At one time I had on approval three eyepieces of the same nominal focal length—usually they are merely pencilled figures on the diaphragm or on a tie-on label, as "$\frac{1}{4}$"" or "$\frac{1}{2}$"", and so on—and found none was what it was supposed to be. As a result one can have very false ideas of one's telescope magnifications.

I have measured the focal lengths of my eyepieces by a method jotted down in my notebook, the origin of which I cannot remember. I am fortunately in the possession of a microscope with a micrometer eyepiece and I dare say many others are in the same position. I calibrated the micrometer many years ago for use with each of my microscope objectives, with the help of a borrowed stage micrometer, so that I can measure minute objects on the stage of the microscope very accurately. In the case considered below the small "object" is the small "image" formed in an astronomical eyepiece. I give the method so that anyone wishing to try it out who has, or can beg, borrow or steal a microscope and micrometer can measure his own focal lengths:

Place a strip of white paper, bearing black vertical lines exactly 10 inches apart, in strong illumination. Place the telescope ocular on the microscope stage, eye end up, and set the microscope at a distance so that the path of light from the strip of paper to the diaphragm of the ocular, via the microscope under-stage mirror, is exactly 100 inches. Then measure the actual size of the small image of the 10-inch strip as seen in the ocular through the microscope. The focal length is the distance the paper is away, multiplied by the size of the image and divided by the distance between the marks on the paper. By choosing the convenient distances mentioned, we have the ocular focal length equal to 10 times the actual size of the small image. For most oculars a low-power (2-inch) microscope objective was found suitable, with higher powers for the shorter focal length oculars. It is a question of obtaining a reasonable size of image to measure.

Another useful adjunct is a Barlow Lens, which is a concave or negative lens, fitted in the draw-tube of a telescope in front of the eyepiece, giving an effective increase in magnification—which can be varied to some extent by altering its distance from the eyepiece—thus enabling one eyepiece to do the work of two or more. Some observers are strong advocates of the Barlow Lens, particularly for use with reflecting telescopes, but while it certainly provides for economy in eyepieces I have not made general use of the one in my possession. In particular, for projecting the Sun I have found that it increases dust shadows, and dust on the eyepieces is a nuisance at any time. I prefer to change my powers by changing eyepieces, as I have a sufficient range.

SUN DIAGONAL. A very important accessory for the solar observer is the Sun diagonal or Herschel Wedge. I emphasize again that the Sun should on no account be looked at direct through a telescope, not even with a dark filter cap on the eyepiece (forget what I said in Chapter Three!), and

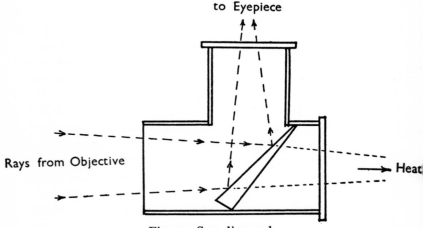

Fig. 21. Sun diagonal.

it is necessary to cut down the excessive light and heat by some means. The sun diagonal is one of the answers, and consists of a fitting for the telescope draw-tube, a partially reflecting surface and a fitting for the eyepiece. Fig. 21 shows such a diagonal in section.

The light passed up through the eyepiece comes from the top surface of the clear glass wedge, giving an image of reduced brilliance, rather like the fainter appearance of lights in a railway carriage when seen from the inside reflected in the windows, while the heat from the Sun, concentrated by the telescope objective or mirror, is passed straight through the open end. The glass is wedge-shaped so that the secondary reflection from its back surface is not also thrown up to the eyepiece to give a double image. Even the reduced light, although not hot, is still much too bright for the eye without the addition of a dark filter cap on the eyepiece, but there is no danger, now, of the filter being cracked. It is convenient to have more than one dark filter to suit the varying brilliance of the Sun and these should be of neutral tint. Sometimes a neutral wedge—from light to dark—is used. This slides in the eyepiece cap so that the strength of the filter can be changed at will.

The Sun as seen through a diagonal fitted to an astronomical (inverting) telescope is reversed E. and W. compared with the naked-eye view. When the diagonal is rotated in the draw-tube the image rotates at twice the rate as shown in Fig. 30 (c) and (d) and when N. is at the top the image is the same as that projected on to a screen behind the eyepiece. See Fig. 30 (e). It is usual to stand at the side of the telescope when using the diagonal, but if you do stand directly behind beware of burning your tie by the heat passing straight through!

OBSERVATORIES. There are mixed views on the question of observatories for amateurs, although it is recognized that some form of shelter is a great advantage, if only as a protection from the wind and consequent vibration. There are many varieties of such shelters in use, from a mere windscreen to a hut with a flat sliding roof, and so on to a fully enclosed building with revolving roof and shuttered opening. The fully open sky view, which is still possible with a roof that slides right off, has the great advantage for the beginners in "star gazing" that they can get to know their way about the constellations. From a narrow open shutter in the roof of an enclosed observatory the view is very limited and much of the beauty of the night sky is lost. However, for solar observation there is every advantage in having a restricted opening to the sky and cutting down as much extraneous light as possible, enabling better and brighter views of the Sun to be obtained.

Amateur astronomers who observe the Sun as only one of the many interesting celestial objects to be studied will not wish to alter any arrangements they already have for a telescope shelter, but for those wishing to take up solar work more seriously it may be possible to modify their existing set-up so that the opening to the sky is limited as far as possible.

The only amateur equipment I can really describe from intimate knowledge is, of course, my own, and in giving a detailed description I only hope there may be found, here and there, a few points of interest, and maybe use, to other amateurs.

MY OBSERVATORY. This is pictured in Plate IX (*a*). It was built in 1933 shortly after I had acquired my last, and present, telescope and when materials were much cheaper than they are today. It was a joint effort by a friendly local builder and myself. I provided the design and scale drawings while he supplied the materials and did most of the hard work. It was very stoutly built, with double walls of matchboarding fixed to a wood frame, the whole being supported on bricks on a concrete base. Incidentally, I had to put wire mesh round the bottom as I soon found that the ventilation space between the wooden floor of the observatory and the concrete base became a favourite haunt for nightly cat gatherings and howlings, particularly in the spring when "the young cat's fancy . . ." etc. In the middle of the base was a separate concrete block going down into the ground some 3 feet as a support for the telescope pillar. The top of this block is a little below floor level to bring the telescope to a convenient height, and the wooden floor has a circular hole round the pillar, with a gap, so that any vibration on the floor is not transmitted to the telescope.

The observatory is octagonal in shape, with a sloping conical roof with four wheels which run on a circular steel ring mounted on the walls. Two shutters are hinged on one side of the roof to open outwards with stays to hold them, while a short handle inside enables the roof to be turned. Actually, the roof is too substantial, making it rather heavy to turn, but like most jobs, I, as the owner, soon developed the necessary knack. The roof is covered with rubberoid. There is a door facing the dwelling house and, on the opposite side, a large window which can be closed by a shutter when desired. Another side has a convenient collapsible shelf and another a closed cabinet for "bits and pieces".

The observatory is wired from the house and has three points. One switches a mains lamp with a swing-over cover to dim, and another goes to a 6-volt bell transformer wired to a two-way switch. One way is connected to a portable 6-volt bulb, through a variable resistance, and the other way switches to the telescope pillar, by bell wires under the floor lino, where I have a rotary switch leading to three small bulbs—one against the R.A. vernier, one against the declination vernier, and one above a clip to take an old watch which I set to sidereal time before trying to find an object with the circles.

For projecting the Sun it is convenient to have the observatory reasonably dim, and I open only one of the roof shutters. I have fixed a roller blind at the top of the opening and another similar blind can be clipped on near the base, each having a dark green holland curtain. A brass frame of soldered bits make a rectangular opening to which the top blind is permanently fixed and this frame runs freely up and down between curtain rails on each side of the shutter opening. By clipping the bottom blind on to the bottom of the brass frame and by manipulating cords I can adjust the rectangular opening to suit the elevation of the telescope when it is directed on the Sun at any time and season. The operation is completed by slipping over the object glass end of the telescope tube a cardboard rectangle (with central hole for telescope) ample to cover the blind opening. The arrangement is shown in Plate XI (*a*).

This gives me quite a dim observatory, and a correspondingly bright projected image is seen. I can darken still further by shuttering the window, but in practice this can be too dark to see when drawing sunspot positions as projected on to a faintly pencilled grid on white card. But the extra darkness is very useful when studying detail on a more magnified, and therefore dimmer, image of a much enlarged portion of the Sun's disk.

MY TELESCOPE. This is shown in Plate IX (*b*). It was purchased second-hand in 1933 and is a fine Cooke 4-inch aperture achromatic refractor of 57 inches focal length (giving a primary image of the Sun of about $\frac{1}{2}$ inch) mounted on an equatorial stand on a cast-iron pillar. It has a weight-driven clock-drive, setting circles and slow motions (by cords) with clamps; also a $1\frac{1}{4}$-inch aperture finder with large field and crosswires. I have since mounted on the body of the main telescope a supplementary painted cardboard tube with a good 2-inch achromatic objective of 20 inches focal length. Although this is only a small aperture it gives a good projected image of the Sun with my 0·3-inch focal length eyepiece, when photographing through the main telescope.

The telescope did not have an adjustable counterweight, but after obtaining a spectroscope, which was rather heavy, I made my own sliding weight by drilling holes (with some trepidation) in the tube of my fine Cooke telescope 'to fix brass brackets which hold the 30-inch long, $\frac{1}{2}$-inch diameter brass rod. A cylindrical weight was cast in a tin with the right amount of lead and a central brass tube which was a sliding fit over the rod, and a clamping screw through the side. This has worked perfectly, enabling me to balance the spectroscope and also my rather lighter camera. It is seen in the photograph which, with that of my observatory, was taken by Patrick Moore.

EYEPIECES. I have a set of eyepieces, most of which are not used very frequently, giving magnifications of 25, 40, 60, 120, 200 and 300. These are all Huygenian and have the Cooke sliding fitting, and I must say that changing eyepieces by sliding is much more convenient and quicker than by the usual unscrewing and screwing. It is a pity there is not a standard sliding fit adopted for all telescope eyepieces. I have an adaptor for standard R.A.S. threaded eyepieces, or accessories, which slides in my draw-tube, and this is useful on occasions. For instance I have a threaded eyepiece which I might almost call a "freak" as it is only 0·127 focal length, giving a magnification on my telescope of 450! I should explain that I had it lent on approval, just for fun, and while it was hopeless on the planets I was surprised to find that when the air was steady I could obtain quite a good, and very highly magnified, projected image of a sunspot. You do not see more detail than with a more moderate power, but it is satisfying to see a sunspot projected as large as your fist.

The lower range of powers is more generally in use, × 60 for projecting a 6-inch image of the Sun's whole disk and × 120 for observing detail and for my solar photography.

As already mentioned, I have a solar spectroscope and a solar camera, and these will be described later.

Chapter Seven

SOLAR SPECTROSCOPE AND
SPECTROHELIOSCOPE

IN CHAPTER FOUR A GENERAL description of the spectrum was given, and prismatic and diffraction spectra are illustrated in the Frontispiece. Reference was made in Chapter Six to the achromatic lens as diagrammatically sketched in Fig. 18 (*b*), and mention was made of an arrangement of prisms whereby the deviation of light rays was corrected while maintaining dispersion. Such a "direct vision" train of three prisms is illustrated in Fig. 22 (*a*) where the outer two prisms are of crown glass and the centre one of flint glass. As the name implies, such an arrangement gives a direct (or straight) view of the resulting spectrum. Being a bright spectrum—with little loss of light through the prisms —such a spectroscope is particularly useful for observing faint objects such as stars. In the case of small star spectroscopes there is no need for a slit, but as the star image is dispersed by the prisms into a very thin line of coloured light (red at one end, violet at the other) it is necessary to widen the spectrum slightly by means of a cylindrical lens which broadens the width only. There are, of course, more elaborate star spectroscopes with slit and eyepiece, or plate holder for photography, with more dispersion for use on the larger telescopes (gathering more light), but the amateur "star gazer" is more likely to have the simplest instrument. I would only add that some of these small 3-prism direct vision spectroscopes are also provided with adjustable slits, but they are really useless for solar work, owing to their limited dispersion.

SOLAR SPECTROSCOPE. There is one form of prism solar

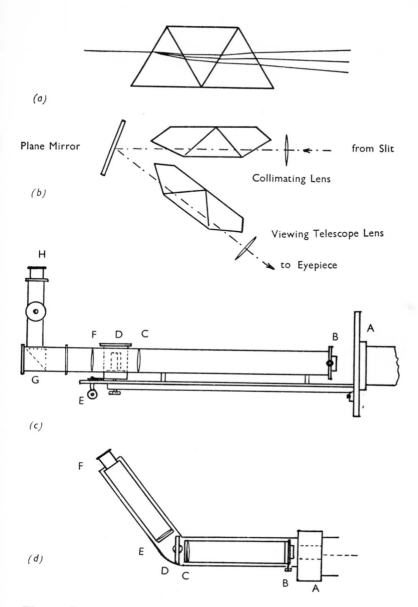

(a)

Plane Mirror

(b)

Collimating Lens

from Slit

Viewing Telescope Lens

to Eyepiece

H

F D C

B A

G

E

(c)

F

E

D C

B A

(d)

Fig. 22. Spectroscopes. (*a*) Direct vision prism train; (*b*) Evershed solar spectroscope; (*c*) Beck-Thorp D.V. diffraction spectroscope; (*d*) Home-constructed diffraction spectroscope.

spectroscope devised by Evershed which should be mentioned, the principle of which is illustrated in Fig. 22 (*b*). This consists of two separate prism trains giving good dispersion, and the different wave-lengths of the spectrum can be brought into view by turning the plane mirror.

Fig. 22 (*c*) shows my own Beck-Thorp Direct Vision Diffraction Prominence Spectroscope, which I will describe in some detail. Most solar spectroscopes today use a diffraction grating, rather than prisms, as it is easier to obtain an extended spectrum and the loss of light is not of importance when observing the Sun. As previously mentioned, such an instrument also spreads the spectrum more evenly—compare the Frontispiece (*a*) and (*b*). I have not yet met anyone with an Evershed type, but Bayley has the "twin" of my Beck instrument.

Referring to Fig. 22 (*c*):

A—is the circular and fixed disk marked with the position angles, 0–360 degrees, permanently attached to a draw-tube to fit my Cooke telescope.

B—the adjustable slit operated by a small screw. The slit was originally a straight one, central with respect to the draw-tube axis, but I realized the advantages of an offset curved slit to more or less fit the curvature of the Sun's limb, so a good friend and precision engineer, S. H. Best, kindly undertook to curve my slit to the correct radius of the image formed by my telescope O.G. He also moved the fixture of the base "girder" on *A* by a distance equal to that radius. The girder rotates round the fixed position angle disk *A* and carries a pointer from which one reads off the degrees. A smaller disk let into *A* (with central hole for rays) is made of black glass, giving a dim reflected view of the solar image positioned on the slit when looked at from the eye end. This enables the curved slit to be set "around"

106

a portion of the Sun's limb and, by rotating the whole instrument round the position circle, the whole circumference of the Sun can be surveyed very easily and quickly. Although the reflection in the black glass is relatively dim, it is still too bright, and I have found it convenient to attach a combined magnifying lens and tinted glass near the eye end of the instrument so that the Sun's image can be seen, and the small figures on the position circle read, in comfort.

C—the collimating lens turning the rays diverging from the slit into parallel rays for passing through the grating which is mounted on one face of a glass prism to give the direct vision.

D—the diffraction grating which is enclosed in a cylindrical metal box and can be turned for adjusting, and be locked by screws at the bottom. The main tube is hinged here so that the shorter rear half can be moved horizontally to scan the whole spectrum.

E—the screw for scanning the whole spectrum. Usually it is kept with the field of view centred on the Hα line.

F—the objective of the viewing telescope.

G—a 45-degree totally reflecting prism to throw the rays from F at right-angles up to the eyepiece.

H—a low-power eyepiece in which the spectrum and the prominences are viewed.

The "elbow" in the viewing telescope is to shorten the whole instrument but, even so, it is 18 inches from A to H. If it had been one straight tube the instrument would have been about 25 inches long and that would have made things impossible for me in my restricted observatory.

As it is, with a length of 18 inches, I am precluded from using the spectroscope during the winter months when the Sun remains low in the sky with the elevation of my telescope correspondingly low. My observatory was built some time before I acquired the spectroscope and I naturally kept down the dimensions. I can have a comfortable view through my telescope in the ordinary way when it is at all angles, but unless it is fairly steeply inclined the spectroscope comes too near my observatory wall to enable me to position my head. Bayley manages with his similar instrument, but another spectroscopic observer, M. Atchison, who used to observe with the 6-inch Cooke telescope of the Hampstead Observatory, tells me that in winter he also was liable to be "impaled" by the eyepiece of the spectroscope and be wedged between the instrument and the observatory wall. Let this be a warning to future observatory builders!

For the above reason I have made myself a small light solar spectroscope similar to the "Sellers" instrument for which full instructions for making are given in a B.A.A. reprint. I have arranged it to fit on my totally reflecting star diagonal so that it operates at right-angles to my telescope and is useful in winter, but let me say at once that the performance is inferior to that of my commercially produced Beck—as is only to be expected.

Fig. 22 (*d*) shows the general idea:

A—fixing on to diagonal, off-centre by radius of Sun.

B—home-made straight adjustable slit; made from thin pieces of brass mounted in a tin cap, with a screw acting against a spring operating on one of the two slit supports, the jaws of the slit being small pieces of razor blade fixed to the thin brass supports with strong adhesive.

C—the collimating lens, 1 inch diameter and 6 inches focal length.

D—a "student's grating", as advertised.

E—the viewing telescope objective, another 1-inch × 6-inch lens.

F—a low-power eyepiece.

The Sellers instrument has a replica grating mounted on a 30-degree prism giving nearly direct vision. Mine has just the replica grating which results in the viewing telescope having to be directed at quite an angle from the straight to see to the end of the spectrum. Also, I have preferred to swivel my viewing half, with the grating as the centre, around a small bolt with locking wing nut, whereas the Sellars arrangement is a solidly fixed one, set on the Hα line specifically for prominence observation. I have preferred to have mine adjustable so that I can have the pleasure of sweeping along the whole length of the spectrum when desired. I do not know if I am alone in deriving pleasure from just seeing the beauty of the whole spectrum.

The tubes are cardboard and a cardboard cover encloses the grating area, the whole being kept reasonably dark. While I have indicated lens sizes, I would not lay down any hard-and-fast rules. Amateurs usually have to make do with any materials they can put their hands on, and slight variations in diameters and focal lengths do not matter anyway. Such home-made spectroscopes can at least show something of the solar prominences, but they should have sufficient dispersion to separate the pair of sodium lines—see Frontispiece (c) and (d).

I will now return to my Beck spectroscope, and in outlining the procedure would stress that this applies to any form of Sun spectroscope likely to be used on an amateur's telescope.

The instrument is pushed into my telescope with the latter in the N. position, as it is rather heavy and the telescope hangs naturally in this position. I have a small screw (a terminal) I can insert in the side with a tapped hole in the telescope tube. This prevents the spectroscope perhaps sliding out a little and destroying the Sun's focus on the slit. By swinging the telescope up into position, while at the same time sliding the adjustable counterweight along the telescope tube, a balance is obtained and the instrument turned on the Sun.

It is easy to project the primary image of the Sun well on to the nearly closed slit, and then the eyepiece of the spectroscope should be accurately focused on the dark Fraunhofer lines. Next it is necessary to focus the Sun's image accurately on the slit. One reads that this should be done by direct examination with a magnifying lens, but I do not understand why it cannot be focused by setting the slit centrally across the Sun's image, and beyond it, so that only half the slit is covered, which is the method I use. The spectrum is then seen to be half bright and half dark across its entire length and the line of demarcation can be sharply focused with the focusing wheel of the telescope.

With everything set, the image of the Sun's limb is directed tangentially to the slit (whether it is straight or curved the procedure is the same) until the bright part of the spectrum tapers off to darkness. This tip of brightness should be just on the Hα line and it is a delicate operation which makes an equatorial stand and clock-drive almost a necessity. The dark line will be seen to turn bright at just the right critical position on the Sun's chromosphere.

There is a great advantage in viewing this area of the spectrum with a red filter on the eyepiece. This cuts out extraneous light, and colours, and sharpens the definition of the line. I use a piece of ruby glass from an old photographic dark lantern which answers admirably. As an aside I must mention the old dodge of cutting glass under water with a pair of scissors. No, I did not believe it either until one day I tried it out surreptitiously "just to see", and it worked, provided the hands were well covered in the bowl of water and my wife did not see what I was doing with her best scissors! After that I was always able to cut glass filters to suit any caps and covers available—I collect these with tubes and tins as a squirrel collects nuts and it is surprising how frequently one finds "just the thing for the job". Do not expect to cut long strips of glass as a tailor does cloth, as it does not open out. But squares are cut with an ordinary glass cutter and the corners can then be trimmed with scissors until a perfect circle is obtained. It is a nice feeling, rather like cutting cheese.

Returning to our procedure, if some bright notches are seen in the dark line it indicates the presence of prominences, perhaps small, perhaps large. The slit is then carefully opened, keeping the bright Sun's limb just off the line, and with luck the actual shape of a prominence will be seen in the widened gap in the spectrum. This is shown in the Frontispiece (e). What may appear as a small bright notch in the nearly closed slit may be a small prominence, but it

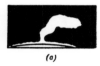 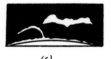

(a)   *(b)*   *(c)*

Fig. 23. Prominence changes in 2 hours—on 1949 May 22. (*a*) at 10.00 hours; (*b*) at 11.00 hours; (*c*) at 12.00 hours (approximately 35,000 miles high).

may be just the "root" of a really large one that expands as the view widens in the open slit.

Fig. 23 is an example of an interesting prominence as recorded in my notebook in 1949, and it shows the rapid change witnessed during two hours.

It is interesting, too, to see the reversal of the blue Hβ hydrogen line, and the chromosphere and any prominences in this colour. The helium line, close to those of sodium in the yellow, also shows bright as an emission line, but the red hydrogen line (preferably seen through a red filter) gives the better definition and is the one in which to see prominences at their best.

If there are any interesting sunspots on the disk the slit

should be trailed across them to see if there are any special features in the dark line spectrum and, in particular, to observe whether there is any reversal of the Hα line to indicate a bright hydrogen flare. Distortions of the line should be looked for, both in prominences and in spots, as they indicate velocities of the gases in the line of sight.

Some sort of scale is useful for drawing and measuring prominences. Bayley used one in his spectroscope eyepiece and I have a glass graticule which slips in my eyepiece when required. This I have calibrated in miles. Atchison had an eyepiece containing a slit wide enough to take the image of the spectroscope slit when the latter is opened for prominence observing. This cuts out all extraneous light and shows up the detail in the lower chromosphere.

The angular position of all prominences round the limb should be noted, and the height and brilliance of any exceptional ones be recorded. Changes in shape over a period, with drawings if possible, should be added, with a note of the dates, times and seeing conditions. Do not expect to see prominences if the sky is at all cloudy or even hazy; the ideal conditions are when the Sun is very clear and steady against a background of clear blue sky.

Finally, do not worry about the dark lines seen along the length of the spectrum when the slit is nearly closed. These are caused by dust particles on the slit and it seems impossible to avoid them, but they disappear when the slit is widened for prominence viewing.

Coming to the spectroheliograph and the spectrohelioscope, already mentioned in Chapter Five, these are fixed telescope arrangements and the Sun's light must be directed on to the telescope objective—which should be of long focus to give a large diameter primary image—by two plane surface-coated mirrors. The first, called a "cœlostat", is clock-driven to direct the Sun's rays continuously on to the second mirror as the Sun passes across the sky. The second mirror

is fixed in front of the telescope objective so that, with the
the first mirror in motion, the rays are continuously directed
down the telescope tube.

It is convenient to have the primary image formed, and
the other parts of the equipment, in a darkened building
with the two mirrors outside directing the rays through a
hole. The mirrors can be suitably protected from the weather
when not in use. I am not going into the details of the mirror
drive, as anyone proposing to construct a spectrohelioscope
would certainly be familiar with the arrangements. I will
only concern myself with the principle of the instruments
themselves, as being of general interest to the amateur solar
observer who does not have, and is unlikely to possess, such
equipment.

SPECTROHELIOGRAPH. This is not used by amateurs, but
is the instrument for photographing features on the Sun in
the restricted light of particular wave-lengths; that is, in the
monochromatic light given out by specific elements in the
Sun's outer layers.

Fig. 24 (a) illustrates the principle. The Sun's image
formed by the long-focus telescope falls on the first slit of a
spectroscope (for clarity the instrument itself is not shown)
and a second slit isolates just one line in the spectrum so
formed. It has already been said that the dark absorption
lines are only dark by comparison, and they admit some
light which can affect a photographic plate. If one imagines
the light passing through such a line it will depict on a plate,
behind the spectrum, an image of the first slit in its monochromatic light. Then, if the first slit is passed across the
solar image while the second slit is passed across the plate,
at *exactly* the same speed, the plate will register an image of
the Sun in the light given out by the particular element in
the Sun's "atmosphere" for which the line is chosen; the
slits can be coupled together mechanically. Alternatively,
the Sun's image can be allowed to cross the first slit while

the plate is moved across the second slit at the same speed, but the first is the usual arrangement.

It is by such an instrument that the photographs in Plate V were taken.

SPECTROHELIOSCOPE.   Here we have an instrument that can be constructed by an amateur, but very few have actually been so made as considerable mechanical skill is required as well as first-class optical parts. The well-known amateurs Newbegin and Sellers each constructed their own instruments, but that of Newbegin is now at the Royal Greenwich Observatory, Herstmonceux, while the Sellers one, bequeathed to the B.A.A., is not in a working condition at the moment, although it can certainly be repaired and brought back into use.

Both of these spectrohelioscopes employed diffraction gratings, but I know of three other amateurs who are seriously considering the construction of instruments of a different pattern, using a train of prisms in place of a grating. Apart from these I am not aware of any spectrohelioscopes in amateur hands in this country, although there are many at professional observatories scattered throughout the world.

First let us consider the general principle as shown in Fig. 24 (*b*) and (*c*). In (*b*) the Sun's image formed by a long-focus objective of, say, 14–18 feet focal length, falls on slit 1. This slit is also in the focus of a concave mirror which directs parallel rays to the diffraction grating where they are dispersed into a spectrum and reflected to a second concave mirror. The spectrum is focused by this on to slit 2, where the image is studied with a positive low-power eyepiece. Now, if slit 1 is made to scan across the Sun's image the focused spectrum will pass across slit 2. If slit 2 is made to move so as to keep *exactly* on the Hα line the eyepiece will show the Sun scanned in the light of hydrogen only, and if the two slits are made to move rapidly backwards and forwards in unison then, by "persistence of vision", an actual hydrogen image of a large area of the Sun will be seen by the observer looking into the eyepiece.

Moreover, if the Hα line shifts slightly to the right or left,

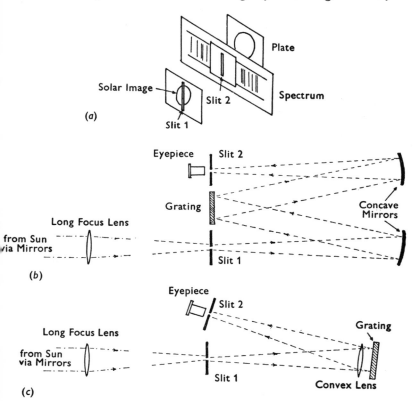

Fig. 24. Spectrohelioscopes. (*a*) Principle of spectroheliograph. (*b*) Spectrohelioscope with 2 concave mirrors. (*c*) Spectrohelioscope with 1 convex lens.

through the Doppler effect of line-of-sight velocities, a device known as a "line shifter" can measure this shift from which can be calculated the speed, towards or away, of the gases in miles per second.

Fig. 24 (*c*) shows an alternative arrangement of one convex lens instead of the two concave mirrors to form the parallel beam on to, and away from, the diffraction grating. The essential feature of any spectrohelioscope is the rapid movement of the two slits working exactly together. The slits can be mechanically connected and electrically driven, or can be actuated by a system of vibrating springs operated electrically, rather like the mechanism of an electric bell. Another, and quite different, method uses two rotating square glass prisms coupled together and scanning the solar image, and the spectrum, across two fixed slits.

Day's spectroscope, as mentioned, employs a train of prisms instead of a grating, and consists of three prisms with a mirror at the back so that the rays are returned through the same prisms giving an extended dispersion through six prisms. I am sure all amateur solar observers will wish him every success with his enterprising effort.

CORONAGRAPH. This was just mentioned in Chapters Four and Five. It is really an instrument for the professional observatories and so outside the province of this account.

It was invented by the French astronomer Lyot as recently as 1930 and, briefly, consists of a telescope and camera with very special lenses of such perfection that they reduce all scattered light to an absolute minimum. Other features also cut down all reflected and diffracted light; in addition most of the coronagraphs are installed in high altitude observatories where there is very little dust in the atmosphere. With an opaque disk exactly to obscure the Sun's image (creating in effect an artificial total eclipse) prominences round the Sun's limb, as well as the inner corona, can be photographed at any time.

MONOCHROMATOR. This is an instrument to take the place of a spectroscope, being a very narrow band filter which just admits the Hα line and which can be used, with a telescope, as an eyepiece, displaying the hydrogen image of the Sun and its appendages. Unfortunately it is not a simple filter that can slip into an eyepiece, but a complicated "sandwich" of sections of quartz and polaroids. In a misguided moment I once obtained from the U.S.A. a leaflet on how to build one's own monochromator, but one reading made it plain that its construction was for the highly skilled worker and not for me. It is a long instrument containing many layers of quartz and polaroids, and it has to be temperature controlled to very fine limits. So I can only continue to look forward to the time when a simple and inexpensive piece of coloured glass, or other material, will be available to cut out all the wave-lengths except the Hα line of the solar spectrum when applied to the amateur's ordinary refracting or reflecting telescope.

In spite of what I have said about the above two instruments not being for the amateur, Arber tells me he is hoping to equip his 4-inch refractor with an attachment so that it can be used as a coronagraph with monochromatic filter for the observation of prominences. But he is in rather a special class as an amateur, with his well-equipped observatory in Manila containing a fine 6-inch refractor, with which he takes his solar photographs, as well as the 4-inch.

Chapter Eight

OBSERVING AND RECORDING

ANYONE, AT ANY TIME, with any telescope, can observe the Sun, on a clear day, with a sun diagonal or by projection, even if he only has a small instrument, but, if projecting out of doors, some shade must be provided so that the direct rays of the Sun, while striking the object glass or mirror, do not fall on the projection card. In the case of a refractor this can conveniently be catered for by slipping a square of cardboard (with central hole) over the object glass end or over the draw-tube at the eye end. See Fig. 5 (a). With a reflector some other form of shading will have to be arranged to suit the particular set-up. With either type of telescope it is preferable to use a simple projection box as described later.

A telescope on a portable, yet steady, stand has the important advantage over a fixed instrument in an observatory in that it can be carried to a sunny spot when the fixed instrument might be shaded by trees or houses, but if such a portable telescope is mounted equatorially it must be placed down in exactly the right direction each time if proper use is to be made of the equatorial movement. If on a rigid tripod the legs must be placed on exactly the correct places each time the telescope is set up for observing. Once the exact three places have been determined they should be marked, and it is useful to have three small metal plates let into the ground for the points of the legs to rest on; one should have a small conical depression made with a twist drill, the second a filed V-slot (facing the direction of the first plate) and the third be plain. By dropping one leg (always the same leg) into the depression, the second will take up its position in the slot, leaving the third to find its own position on the plain plate. This does not prevent the telescope being planted at any convenient spot in a garden for the casual observation when accuracy of position is unimportant.

I cannot do better than briefly outline the programme of solar observation undertaken by the Solar Observing Section of the British Astronomical Association (B.A.A.) under the guidance of its Director—at present H. Hill:

(a) Look for, and note position of, any naked-eye sunspots.

(b) Make daily disk drawings, showing all visible features.

(c) Draw individual interesting and active spots.

(d) Count number of Active Areas.

(e) Note and draw any spots showing the Wilson Effect.

(f) Note any veiled (indefined) spots.

(g) Note any pronounced faculæ.

(h) Also—make spectroscopic and photographic records.

The essence of this work is co-operation, and when one considers the vagaries of climate it will readily be appreciated that a fairly complete daily record can only be obtained from a number of workers, many in different parts of the world recording the Sun's features, each on his clear days. For instance, in the Director's annual report on solar activity during the year 1960 the members of the Section, as a whole, only missed five days, while in 1961 not a single day was missed; an excellent record. In contrast I, as only one of the observers, made drawings on 202 days in 1960 and on 213 days in 1961 which are, I suppose, fair averages for one observer in this country.

Coming to the methods of observing and recording, I can only give in full detail my own practice, but the same principles will apply to almost any telescope. I depend on the projection method, but frequently have a preliminary survey of the Sun through the sun diagonal. I have found, however, that I cannot see such small detail by this method as by projection, and small pores are likely to be overlooked. If a really large sunspot is seen in the diagonal this does prompt me to look through a filter with the naked eye to see if a "naked-eye spot" can be seen and recorded. It should be noted that the diagonal view is reversed east and west compared with that seen with the naked eye, but it is the same as a projected image.

With only one of the observatory roof shutters open, I set the telescope in the direction of the Sun, guided by the shadow of its tube, then pull down the top spring blind with its rectangular brass strip frame on curtain runners, clip on the lower blind to the base of the shutter opening, raise it to hook on to the bottom of the frame and adjust the height of the opening by cords. The projection fitting and card to take the image are attached to the eye end of the telescope, which is then accurately directed to the Sun and the rectangular piece of cardboard slipped over the object glass well covering the gap between the two blinds.

The Sun's disk projected on the card is shown in Plate X (*a*), which is another photograph taken by Patrick Moore.

The actual support for my projection card is sketched in Fig. 25:

A—A thin brass strip, felt-lined for smooth rotation round the telescope tube.

B—A wood block shaped to fit the brass strip and telescope tube, firmly glued and screwed to the brass strip. It contains a brass tube which takes, as a sliding fit, a round $\frac{3}{8}$-inch aluminium rod which is clamped in position with a wood screw.

C—The aluminium rod, 17 inches long.

D—A double carrier (wood) for the projection card frame, the front half having a wood-screw for fixing at about the correct position (marked by two nicks in the rod, indicating maximum and minimum distances—for our summer and winter sun diameters). This is coupled to

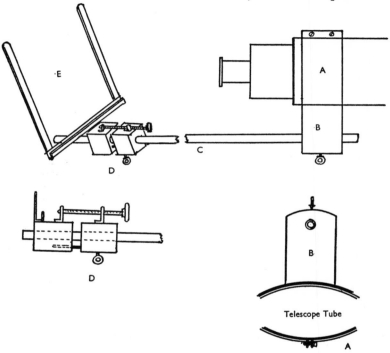

Fig. 25. Projecting arrangements.

the back half by a quick-motion screw with knob (a radio part) and has near the bottom a fixed pin which slides in a corresponding hole in the back half to maintain alignment. This enables a quick adjustment to be made to fit the Sun's image *exactly* to the 6-inch circle drawn on the projection card. Each half contains a short brass tube in which the rod slides easily. By the way, tightening screws such as in *B* and *D* are very conveniently made by soldering thin metal washers into the slots of wood-screws. The normal distance from eye-piece to card is about 10 inches.

E—A brass strip frame (bits soldered together) to support the card.

The Sun's (out of focus) primary image is first centred on the card by turning *E* on the rod and clamping *D*. After that any rotation of the card for orientating the image is done by twisting the whole round the telescope tube at *A*, the telescope tube being the axis of rotation.

The clock-drive is started, the motions clamped, my × 60 eyepiece inserted and the Sun's 6-inch image focused on the card. Slight adjustment is made by the quick-motion screw on *D* until the refocused image exactly fits the drawn circle.

The card must now be orientated correctly. If there is a spot visible on the disk, as there usually is, it is brought to one of the horizontal E/W lines on the card by the declination slow motion and moved backwards and forwards by the R.A. slow motion until, by turning the whole arrangement round the telescope tube, the card is truly orientated E. and W., the spot accurately following the line. If there is no spot immediately visible then the N. or S. limb of the Sun must be made to travel along the line.

The projection card has mounted on its face a piece of matt double-weight bromide photographic paper, unexposed and fixed in hypo, which I have found to have a very satisfactory surface for the purpose. It is smooth and does not have any shine to give unwanted reflections. The paper can conveniently be mounted on a stiff card by photo corners, allowing easy replacement when "grubby" as it does collect some dirt spots in the course of time and these might be taken for sunspots! Actually, of course, solar markings can easily be distinguished from dust by slightly moving the image, but one does want as clean a card as possible.

The projected image has N. at the top, S. at the bottom, E. to the right and W. to the left and is like a naked-eye view seen in a mirror. If there is no telescope-drive the image will drift to the left. Also, sunspots will appear at the right limb and be carried from right to left by the Sun's rotation as the days go by, disappearing round the left limb.

Perhaps my setting up sounds a long job in the way I have described it, but the fact is that it takes about two minutes from going into my observatory to the time when I commence my drawing.

The projection card bears a 6-inch circle divided by fine pencil lines into $\frac{1}{2}$-inch squares with diagonals; fine and faint enough so as not to hide any small sunspots. This is illustrated on a small scale in Fig. 26 (a). It will be understood that this drawing had to be too heavily lined for reproduction reasons. A stout card, mounted on hardboard for handling and with a white surface, has a similar 6-inch circle and grid, but in this case the lines are heavily inked. Blank forms on thin typewriting copy paper ($8'' \times 6\frac{1}{2}''$) are used for the recordings and each time a drawing is completed a new blank is clipped over the hardboard card with the circles and compass points coinciding, ready for the next recording. The blank forms are quickly prepared with the help of a 6-inch protractor and I make a supply for stock whenever I feel like it. The paper is thin enough for the heavily inked lines underneath to be easily visible.

The stout card and a blank form are shown in Fig. 26 (b) and (c) respectively, the blank already bearing the circle, the N., S., E. and W. points and the centre cross, the instrument used and the observer's name. Fig. 26 (d) shows the completed drawing as sent to the Director of the Section.

The clock-drive is a great asset, as the Sun's image remains on the card in its correct orientation (even while

temporary clouds obscure the Sun) and only a slight adjustment in R.A. is necessary from time to time to keep it exactly
coincident with the drawn circle. Without a drive the image
must, of course, be continually brought back by hand before
each spot position is recorded.

With an altazimuth stand it is necessary to make frequent
changes in the orientation of the projection card so as to
keep the Sun's movement along the E/W line, checking
periodically by letting a spot trail. The correction should be
made at least every five minutes—if the observation takes
longer than this—to ensure reasonable accuracy in the completed drawing.

Everything being set, any spots that are visible are seen
on the faint pencilled grid and then exact positions can
easily and rapidly be copied on to the thin paper guided by
the similar grid under the paper. A pencilled note of the
time (U.T.) and seeing conditions is added.

A higher power eyepiece is now substituted—usually my
x 120 giving a solar diameter of 12 inches (with the card
left as for the 6-inch disk) or my x 200 giving a 20-inch Sun.
By slowly scanning the whole of the Sun's enlarged face,
bit by bit in R.A. and declination, one can frequently detect
small pores that were originally missed but which, knowing
where to look, can then be found on the 6-inch image and
their positions recorded. With these higher magnifications,
and when the air is really steady, the granulation of the Sun's
surface is well seen and the more interesting sunspots can
be drawn on a larger scale. For the latter it is useful to clip
a piece of thin plain paper over the stout card and project
the enlarged spot, or group, on to the pencilled card. The grid
is then a useful reference for delineating the shapes in detail.

A note is made of any striking faculæ, particularly any
small bright patches near the poles, and these should be
indicated on the drawing by dotted lines. Spotless days
should always be recorded as such.

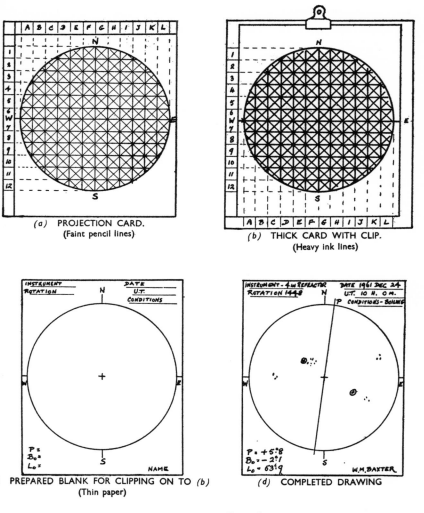

Fig. 26. Spot recording charts.

The actual daily drawing at the telescope is made in a matter of minutes and very often advantage can be taken of quite small breaks in clouds. It is almost a matter of "honour" for a keen solar observer to obtain his daily drawing whenever the Sun appears, even to the extent of sacrificing a breakfast!

Plate X (*b*) shows the projected image on a large white card fixed to the inside of the sloping roof of the observatory on the opposite side to the shutter. For this I attach a small 1½-inch diameter silver-surface mirror (Government surplus, at the cost of 1*s.* 6*d.*) on a universal joint mounted in *B* of Fig. 25 and just behind the eyepiece. With this and a high power eyepiece, and given steady air, a really large projected image can be obtained and it can be seen by several persons to advantage, looking very striking indeed, but now the image appears with N. at the bottom, S. at the top, E. to the right and W. to the left.

Coming indoors, and at any convenient time, the blank with spot positions marked, is completed by filling in the remaining details:

Instrument. This can already be on the prepared blank if the same instrument is always used—as it should be for true comparisons to be made.

Rotation Number. The Sun's rotation period (mean) is taken as 25·38 days and rotations are numbered in sequence. At the beginning of 1961 the rotation number was 1436 and at the end of that year No. 1449. The rotation numbers, with the date and time of the commencement of each, are given in the *B.A.A. Handbook* for each year.

Date. This should be given with the year first, followed by the month and then the day.

Time. Universal Time should always be recorded (U.T.). This is Greenwich Mean Time counted from midnight, i.e. our ordinary time of day in this country, but not in other countries. Care must be taken to see that one hour is deducted from clock time if "Summer Time" is in force.

P, B_0, L_0. These are the Sun's co-ordinates for the time the drawing is made, being the tilt of the Sun's axis, the latitude of the centre of the Sun's disk and the longitude of the central meridian respectively, all in degrees.

P is the position angle of the N. end of the axis of rotation, with + if E. and − if W.

B_0 is the latitude of the centre of the Sun's disk, + or −, indicating the tilt of the axis towards or away from us.

L_0 is the longitude of the centre of the disk or of the central meridian.

Tables giving the three values for noon, at 4-day intervals, are printed in the *B.A.A. Handbook* for the year, and the figures for intermediate dates and times are easily interpolated. The Longitude (L_0) decreases with time by 13·2 degrees a day, and a table is given to adjust the noon figure to that of the time of observation. This is all made clear in the *Handbook*.*

Having noted the figures, they are entered on the drawing and the Sun's axis drawn to correspond with the angle of P.

Conditions. It only remains to enter this item, already pencilled in at the actual time of the observation. The kind of remarks to enter are:

> Clear
> Boiling
> Haze
> Thin Fog
> Thin Cloud
> Drifting Clouds
> Wind Vibration

By far the most frequent condition met with on a clear sunny day—in this country at least—is "Boiling", which

* See Appendix for an example.

means that our atmosphere is very disturbed, causing ripples over the Sun's image and a wavy limb. Spots go in and out of focus as you watch, and the smaller spots become invisible. Sometimes I enter "Boiling Violently", and these are obviously not the times to attempt solar photography.

The most settled time of the day seems to be in the morning when the Sun is at an elevation of about 30 degrees or so above the horizon, as when it is low it suffers from air currents rising from the ground and buildings, and when high —at least in summer—the air gets very hot, again giving rise to excessive atmospheric turbulence. The sharpest definition is often found on misty days, and in England there is probably nothing better than a November day with a very thin mist or light fog. On such a day, with the air perfectly still, I have had the thrill of seeing a large projected image of the Sun, perfectly steady with the smallest details, including surface granulation, visible rather like a steel engraving. Such occasions are unfortunately rare, but when they do appear the most should be made of them, particularly if there are some interesting features to photograph.

In the drawing as completed—see Fig. 26 (d)—it will be seen that the E. and W. points are reversed mirror-wise when compared with the naked-eye view or with photographs, so that if it is desired for any reason to compare a drawing with naked-eye spots, or with a photograph, the drawing must be looked at through the back (being on thin paper this can be done) or in a mirror. Orientations, although somewhat confusing, are not important provided they are clearly defined.

Personally, I always make duplicate copies of my finished disk drawings and this is easily done by putting another "blank" over the drawing, so that the circles and points coincide, and holding the pair against a window. It is only a minute's job to trace any spots, mark the axis and complete the information on the duplicate. This enables me to keep a copy for myself while sending the "original" to the B.A.A.

Solar Section Director who, in due course, will produce a report on solar activity for the year. It is then useful to be able to refer back to one's own drawings when reading of the various activities of the Sun as reported collectively by a large number of observers. Reports appear elsewhere too, from time to time, and it is always interesting to turn up personal records for comparison purposes.

Copies are also made, when requested, for the Radio Astronomy Section of the B.A.A. during the period each year when the Sun passes in front of the Crab Nebula, which is a strong source of radio emission. This phenomenon enables the radio astronomers to study the extent of the Sun's corona by its effect on the radio waves, and daily drawings of sunspot activity tell them "what is going on" visually at the time. For further information on this subject I must again refer my readers to other books, but I have already mentioned the possibility of the Earth being within the extensions of the Sun's corona.

Active Areas are counted from the drawing. Every spot, however small, counts as a separate active area if it is at least 10 degrees of latitude or longitude from its nearest neighbour. A large group, however, spread out in area, is still one active area unless it has distinct separate centres of activity at least 10 degrees apart. One might think it would often be difficult to decide, in the case of a much-spotted Sun, how many separate areas were visible, but in practice this difficulty rarely arises, as more often than not the spots fall into separate groupings.

A card marked as Fig. 27 (a) is a useful guide for anticipating the position of a spot when the Sun has been obscured by clouds for a few days. It shows "daily" lines of longitude and by counting days—towards the left if the projection drawing is laid on it with the Sun's axis (P) coincident with the central meridian line—one can readily see when the spot will reach any particular position such as the central meridian or the West limb. By allowing about fourteen days from its disappearance round this limb, one can estimate, of course, when *and where* a long-lived spot is due to reappear at the East limb.

On the back of the same card I have marked 10 degree differences as a guide as to whether a pair are separate areas

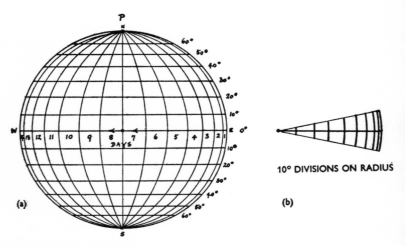

10° DIVISIONS ON RADIUS

(a) (b)

14 DIVISIONS ON HEMISPHERE REPRESENTING DAYS

Fig. 27. Sunspot positions. (*a*) For anticipating spot positions; (*b*) for measuring 10-degree distances.

or not—Fig. 27 (*b*). It is not always easy to estimate a distance of 10 degrees, particularly towards the limb where foreshortening has to be allowed for. In order to measure this the drawing is placed on the card with the centre of the Sun's disc at the apex of the triangle, and the nearest spot (*x*) on the centre line. See Fig. 28 (*a*). The drawing is now pivoted on this spot so that the second spot (*y*) also appears on the centre line and the distance between the two can then readily be seen. See Fig. 28 (*b*).

The card is easily prepared by dividing the top (or bottom) semicircle of Fig. 27 (*a*) into fourteen parts and dropping perpendiculars to the centre line, while for (*b*) the

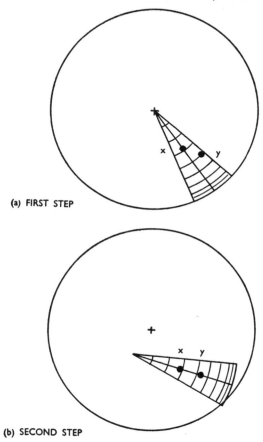

(a) FIRST STEP

(b) SECOND STEP

Fig. 28. Angular distance between spots.
Using the sector shown in Fig. 27 (*b*).

quarter circumference is divided into nine parts, only a small sector being required.

The drawings are sent to the Director at least at the

completion of every rotation, or monthly with a note of the number of Active Areas and the Mean Daily Frequency of the latter, being, of course, the total for the month divided by the number of observing days. These M.D.F. figures are also entered on a form provided by the Director which is returned to him, filled in, at the end of each year. Again, I like to keep a duplicate of this record for my own information.

The Director maintains a continuous record of solar activity from all observers' contributions, but I keep my own graph of M.D.F. figures and can follow the sunspot cycle with its rise and fall in activity over the years—rather like Fig. 9, but I shall not live to see so many cycles as are shown there! The Director goes further and not only prepares his graph from the collective information received from many observers but he plots his own "butterfly diagram" like Fig. 11.

There are other systems of recording sunspot activity, the most important being that used by the Swiss Federal Observatory, Zürich, under the direction of Professor M. Waldmeier, known as Wolf sunspot relative numbers, or Zürich numbers (instituted by Wolf as long ago as 1848). In this system the number of component spots is noted as g and is multiplied by a factor of 10, while the total number of individual spots is noted as f. A multiplying factor k is given to each observer, depending on his instrument and observing ability, so that the observations of all taking part—at present some thirty professional and about the same number of amateur observers in various parts of the world—are directly comparable.

Sunspot numbers R is given by:

$$R = k \, (10g + f)$$

For example, if there are two groups each consisting of two spots and, in addition, one isolated spot: $g = 3$ and $f = 5$. Then

$$R = 10g + f = 35 \text{ multiplied by the observer's } k.$$

It sounds very arbitrary, but it has been found to work very well indeed since it was instituted, and the system is still continued.

The Royal Greenwich Observatory records sunspot activity by the total area of all spots, including umbra and penumbra, as determined from photographs.

Observers who do not have an observatory that can be somewhat darkened, or who observe out of doors, can obtain the same results by projecting the solar image into a dark box fitted to the eyepiece of their telescopes. This can be a simple cardboard box with half of one side open for viewing, one end fitting on the telescope and the other end containing the projection card large enough to cover a 6-inch diameter image. Such a box should have the front end reinforced with a square of sheet brass or tin cemented to the cardboard. A circular hole is made through both, the edge of which is gripped by the screwed eyepiece when the latter is tightened in the draw-tube. This will support the box without sagging. Alternatively the box can be simply a light wood frame covered with black paper; the inside should be blackened anyway.

Fig. 29 (*a*) and (*b*) are sketches of more substantial projection boxes, (*a*) being a balsa wood box with glued joints constructed by Joy for his 10-inch reflector stopped down to 4 inches, while (*b*) is a metal drum used by Bayley on his 5¾-inch refractor. I am indebted to these two amateurs for providing me with particulars.

In Joy's arrangement the projection card is really in the form of a box lid which can slide to allow for the variation in the apparent diameter of the Sun at different times of the year—it varies between 31½ and 32½ minutes of arc and must be allowed for.

Bayley slides his drum on the draw-tube to adjust for a 6-inch image diameter at the date of observation. The draw tube has a zero mark and he adjusts from this with a millimetre rule from readings on a graph against different dates throughout the year. Another graph gives him readings for a set diameter with a high-power eyepiece when only a portion of the disk is projected. The ends of his drum are stiffened by ribs and the back one is slightly dished outwards so that the projection card (fixed to a wooden turntable with centre bolt passing through the end and tightened with a wing nut) is held fast once the turntable has been set for the correct orientation and clamped. His arrangement allows for the projection box to be turned to the most convenient viewing position and the card to be independently orientated.

Addey had a projection box very similar to Fig. 29 (a) on his 4-inch equatorial reflector with a suitable counterpoise fixed to the opposite side of his telescope tube, which is of square section and of wood. Both he and Bayley found it useful to cover their heads and projection boxes with a dark cloth rather like that once used by photographers. This gives them a very bright image, by contrast, and shows up small detail that might otherwise be missed. I was certainly very impressed with the projected Sun as seen with Addey's equipment, and I am sure any newcomer to solar work would soon display enough ingenuity to make some suitable "contraption" to suit his own telescope.

It is sometimes necessary to define the position of a spot or other centre of activity on the Sun's disk by latitude and longitude. What are known as Stonyhurst Disks are very convenient for this purpose, being a series of eight semi-transparent prints of the solar disk with the lines of latitude and longitude marked. The different prints are used for different dates, varying with the tilt of the Sun's axis from +7 to −7 degrees (towards and away from us). The Sun's image can be projected direct on to the selected print and the latitude and longitude of any spots noted at once. Alternatively, the prints can be placed over the drawings and the co-ordinates read off. The true longitude is quickly

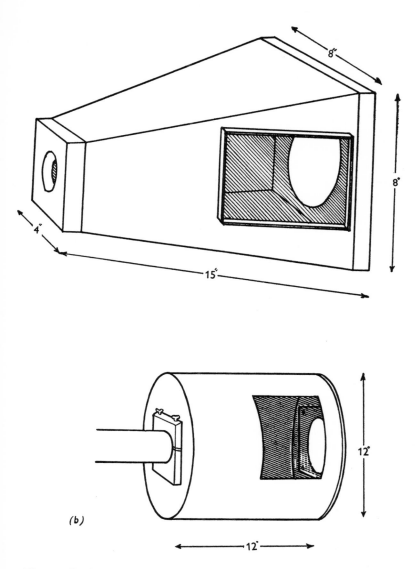

Fig 29. Projection boxes. (a) Of balsa wood; (b) a metal drum.

calculated from the longitude of the central meridian of the day (L_0) as given in the *Handbook*.

Unfortunately, the set of prints is rather expensive, but the same results can be obtained, with little extra trouble, by using a single "Porter's Disk" obtainable from the B.A.A. for a few pence. The disk drawing on thin paper is placed over the Porter's Disk card and the position of the spot noted against horizontal, vertical and concentric lines printed on the face of the card, the back of which contains full directions for the simple calculations required.

Another measurement which is sometimes desired is the area of a sunspot or of a group. Areas are defined as being so many millionths of the Sun's hemispherical (visible) surface and a convenient way of measuring is given in Sidgwick's *Observational Astronomy for Amateurs* using 1-mm. graph paper squares for a solar diameter of 24 inches. The spots can be projected on, or drawn on, the graph paper, but I find a transparent (photographic) copy on glass of the graph squares very useful for measuring my sunspot photographs, as I normally enlarge to a 24-inch sun diameter anyway. Tracing paper with 1-mm. squares is obtainable and can be used in the same way.

The number of small squares covered by the spot, or group, are counted and multiplied by 1·72 to give the area in millionths of the Sun's hemisphere. A correction has to be made for foreshortening, unless the spot is in the centre of the disk, and this is conveniently calculated from the 6-inch disk drawings, the distance of the spot from the centre being measured in inches. The table on p. 137 (from Sidgwick) gives the multiplier for different distances.

An area can only be measured approximately in any case, as the perimeters of spots are not so clearly defined that the small squares can accurately be counted—half-squares are not counted but larger than half squares are counted as whole squares. It can be a rather tedious, if simple, exercise,

Distance (ins.)	Multiplier	Distance (ins.)	Multiplier
0·42	1·01	1·80	1·25
0·59	1·02	1·92	1·30
0·72	1·03	2·02	1·35
0·82	1.04	2·10	1·40
0·91	1·05	2·17	1·45
0·99	1·06	2·24	1·50
1·07	1·07	2·34	1·60
1·13	1·08	2·43	1·70
1·19	1·09	2·50	1·80
1·25	1·10	2·55	1·90
1·48	1·15	2·60	2·00
1·66	1.20		

but the results are often impressive in their magnitude.

Rather as an aside, I will mention transits of bodies in front of the Sun, which are not "solar" phenomena, of course. Mercury is a small planet which very occasionally passes in front of the Sun, as seen from the Earth, during its orbital journey. If it passes exactly in front of the Sun it appears as a small black dot. It is interesting for the solar observer to compare the intensity of its blackness with the perhaps lighter shade of the dark umbra of any sunspot that may be present. I observed a recent transit, in November 1960, in the company of another regular solar observer, C. J. Aston, at London, S.W.15, with his 3¼-inch altazimuth refractor and × 160 eyepiece, as I was unable to see the Sun from my observatory at that month and hour owing to obstruction by houses. I certainly had the impression that the umbra of a near-by spot was not so intensely black as Mercury appeared to be.

Transits occur in pairs, and the next transits of Mercury were due in 1970 and 1973.

Transits of Venus are much more impressive, as this planet is so much larger than Mercury, and nearer, but the last transit was in 1882, before I was born, and the next two will not be until 2004 and 2012, so that I am sure I shall not be here to witness them.

There is another type of "transit" which can be seen more frequently and, while of no scientific value, is not without some interest to the Sun watcher. I refer to the passage of birds in front of the Sun, very often in arrow-head formation. Even an aircraft holds attention as it passes across the Sun's face projected on the card, with its black silhouette looking like one of those identification chart outlines. The hot gases in the wake of a "jet" are impressive too, with black whirls of turbulence created in the atmosphere entirely upsetting the sharpness of the Sun's image. Drifting clouds often show the different directions of air currents at different levels, with one layer moving one way and another moving, at the same time, perhaps at right-angles to the first.

Observers with spectroscopes will turn their instruments on the Sun whenever conditions are favourable and, with the slightly widened slit placed tangential to the Sun's limb, and the spectrum centred on the Hα line, search the circumference for prominences. Then, with the slit wider still, they will record the shapes, heights, brilliance and position angles of any so found. If an active type of prominence is seen it should be sketched to scale and, if possible, this should be repeated at intervals so that any changes that occur can be recorded.

Good seeing conditions are very necessary for clearly observing prominences, and the ideal day is one when the atmosphere is clear and steady, with the sky around the Sun free of all haze. The Hα line should also be scanned across any active sunspot groups that may be present to see if there is any distortion of the line due to line-of-sight velocities of the gases, and if there is any sign of a flare.

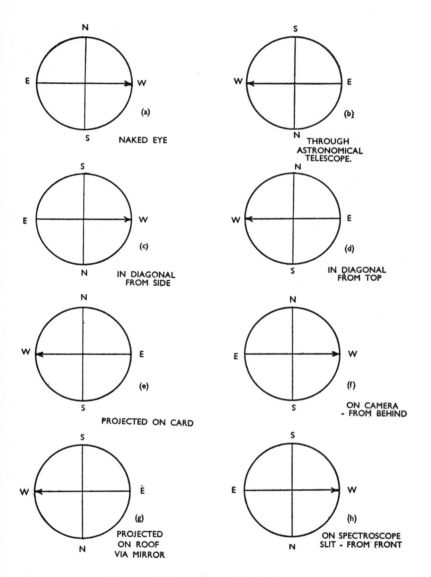

Fig. 30. Orientation of Sun's image.

I have no personal experience with a spectrohelioscope, but the fortunate possessors of such instruments will undoubtedly be able to occupy many happy hours whenever the Sun is clear. They will look for hydrogen phenomena on the disk itself and make drawings of all they see, noting the many changes that take place, including velocities in the line of sight.

It will be useful, perhaps, to refer again to the question of the orientation of the Sun's image, already mentioned several times. There are so many possible optical arrangements with various instruments, refractors and reflectors, with multiple reflections, that the only sure way to ascertain what one is actually seeing is to note which way the Sun drifts across the field of view with the telescope stationary. This will be towards the west. Then by tilting the front end of the telescope slightly upwards the north limb of the Sun (south limb for southern hemisphere observers) will be brought into view and the four points will be fixed. Orientation in a spectroscope and in a spectrohelioscope can be particularly confusing. However, I have sketched Fig. 30 to show just some of the more usual orientations encountered with different arrangements.

In concluding this chapter I am again conscious that my readers may think I have gone into details at too great a length, but I hasten to assure them that the whole procedure of observing and recording is really very simple and is easily picked up by any beginner in the solar field of astronomy. While I have stressed co-operation I would like to make it clear that it is not necessary for an amateur to belong to an association of observers to experience the pleasure of keeping a watch on the Sun. Maintaining one's own personal records can be a real source of interest.

I have already referred to Addey's completely home-made (including the mirror) 4-inch equatorial reflecting telescope, and I would like to finish with an extract from a letter I once received from him. His words should be an encouragement to all beginners with limited resources:

My actual observations have nothing remarkable about them, but looking back over the sixteen years during which I have managed to make observations on most days, I feel that what I have been able to do should encourage any amateur who has only a small telescope available. The combining of the observations has given me a much better picture of the variations in solar activity than I could have obtained from books alone. Although I have not made any remarkable discoveries, I have noticed and recorded phenomena which only subsequently were brought to my notice from other sources. Such "discoveries" give just as much pleasure as if they were unique.

Chapter Nine

SOLAR PHOTOGRAPHY

ALTHOUGH THIS IS THE BRANCH of work in which I am particularly interested, and the plates at the back of this book are all examples of my solar photography, I am reluctant to suggest any "rules", as instruments and conditions vary so much that it is really a subject for individual experiment. I can only indicate the methods used by one or two other workers in this field and give details of the procedures that I have myself adopted after much trial and error in the past.

Most amateurs who regularly observe the Sun will have a sun diagonal (Fig. 21) and a first experiment can be made by photographing with an ordinary small hand camera held to the eyepiece on the diagonal. A dark filter should be on the camera lens, or on the eyepiece perhaps dropped into the recess of the latter. The filter should be rather a lighter shade than the one generally used visually, the camera stop should be open and the focus set to infinity. The telescope is focused visually through the filter in the ordinary way, using a supplementary neutral filter if necessary.

The camera, with the light shade of filter, is then fixed to, or just held against, the eyepiece, and a number of short exposures made as a test, say at 1/50, 1/100 and 1/250 sec. depending on the range available. By this means a number of guiding exposures can be made on one film at little expense and, although the resulting image will be small, some interesting results can be obtained, particularly if a spot group is photographed over a number of days to show the movement across the disk as the Sun rotates and the changes in its shape. Plate VII (*a*) is an example of a photograph taken in this way with my hand camera, with an exposure of 1/50 second.

If one is the possessor of an observatory that can be reasonably darkened, and drawings are customarily made by projecting the Sun, it will immediately be obvious that the projected image can be photographed with an ordinary camera. The difficulty here, apart from securing the dark surroundings, is to fix the camera near enough to the telescope axis to avoid distortion of the projected image which, of course, appears more and more elliptical as the camera moves away from the axis. However, by having the camera as close as possible to the telescope and getting as near to the image as the camera focusing will allow without the eyepiece being in the way, and slightly tilting the projection card (but not enough to throw the sides of the image out of focus), this distortion can largely be eliminated. Plate VII (b) was taken in my darkened observatory in this way, but I am afraid the Sun is not quite circular.

In this case the exposure will be longer than that through the diagonal and the camera should be on a stand. An ordinary meter will give a good guide for exposure but, again, it means trial and error.

Anyone observing in the open might be able to rig up some sort of dark cloth covering for photographing in this way, but it would have completely to cover the eye end of the telescope, the camera, the projection card and even the photographer unless the camera was one that could be preset; all of which would not be easy, but not impossible.

Arber, of Manila, gave an account of his methods of solar photography to a B.A.A. meeting, from which it appeared that he projected an 8-inch disk for drawing and sunspot counts, but for photography he used a 2-inch eyepiece on his 6-inch O.G. refractor, giving a solar image on the film of 104 mm. (about 4 inches). For taking early morning photographs, when conditions are likely to be at their best, he uses the full 6 inches of his object glass, but near midday he stops down to 4 inches, or even 3 inches, to reduce the excessive heat. The 104 mm. disk can be accommodated on the 5″ × 7″ Super Kodalith sheet films that he uses. The fastest shutter speed is about 1/200 sec. and the shutter is released through an air bulb. He used a 2-inch diameter orange filter combined with a neutral filter of the same diameter, both being about 10 inches in front of the eyepiece, the filters reducing the light by about 4 times.

A 3-inch guide telescope gives a sun diameter of 1 inch projected on to a screen with cross-hairs, and this enables him to avoid exposing during passing clouds or when the image is very unsteady. He usually takes two pictures of any interesting sunspots.

The films can safely be developed in a dark red light, and he uses Kodak D-11 developer diluted 1:5. Plate IV is a particularly fine photograph of the Sun's disk taken by Arber on 1957 December 25 during sunspot "maximum".

H. N. D. Wright, of London, S.W.17, is a solar photographer who has very simple equipment such as is available to most serious amateur astronomers. He works with a simple 3-inch equatorial refractor with an ordinary folding camera (minus lens, of course) mounted at the eye end. He originally experimented, not very successfully, with 120 films, but after hearing Arber's methods he obtained a ¼-plate holder with adaptor for sheet film and fitted it to his camera. He, too, has adopted Kodalith Ortho Type 3 film, which has a fine grain, is slow and can be developed in a red light.

The camera has a Compur-type shutter with a range of speeds, and so far he has estimated his exposures, but is now experimenting with an exposure meter held against the focusing screen. He uses a 28 mm. × 4 orange filter in combination with a heat filter of optical quality, fitted behind the shutter, and with a × 132 eyepiece he secures an image on the film equal to a solar diameter of 5·6 inches.

His finder telescope projects a small image of the Sun. This is only ¼ inch diameter and has to be strongly filtered to cut down glare. Part of a magnetic calendar plate is used with the index ring carrying the projection screen bearing an inscribed circle of the required size. With the part of the Sun to be photographed centred on the camera focusing screen the index ring is placed so that the finder image is on the inscribed circle. The telescope is then set about half a minute of time in advance of the Sun so that its drift carries it to the correct position, allowing any vibration transmitted to the telescope or stand to damp out. A long cable release helps too.

Plate VIII (*a*) shows Wright's set-up, and (*b*) and (*c*) are two of his solar photographs taken as described. He has been particularly successful in capturing the surface granulation in (*c*).

J. B. Orr, of Auckland, New Zealand, is another amateur astronomer who, like Arber, has an exceptionally well-equipped observatory containing a 12-inch Maksutov reflector as well as a 4-inch refractor with which he takes solar photographs. This 4-inch has a first-class objective by Goto, and Orr reduces the Sun's light for photography by means of a large optical filter fitted over the telescope object glass. This is perhaps an expensive way of doing it, but the results have shown that it is a very efficient way and it certainly overcomes all heat problems, as the filter only receives normal sun heat. This rather special filter was obtained from the U.S.A. and is advertised there as having a thin coat of aluminium on one side to reflect the Sun's heat and excess light away, enabling the telescope to be turned on the Sun for visual observations, or for photography, in the same way as one would look at the Moon. Orr says it transmits strongly in the orange part of the spectrum, and he has certainly taken very successful disk photographs, as well as some on a larger scale with a $\frac{1}{2}$-inch eyepiece. The solar camera is a commercial one supplied by Goto.

Orr uses Ilford R 40 $\frac{1}{4}$-plates with an exposure time of 1/25 sec. with his 40-mm. eyepiece and $\frac{1}{2}$-sec. with his $\frac{1}{2}$-inch eyepiece. The disk size with the former is $2\frac{7}{8}$ inches and with the latter $10\frac{1}{2}$ inches. The camera has a Compur shutter which opens to 1 inch diameter and has speeds from 1/5 to 1/200 sec., and exposure times are estimated, differences being compensated to some extent by varying the development times.

As the O.G. filter eliminates the blue rays a correcting filter on the eyepiece in unnecessary. Focusing is done with a magnifier on the ground-glass screen and Orr feels that this is superior to having a clear glass patch with a cross to focus on. He has discarded his projected image from a 2-inch guide telescope and now relies on his telescope setting and electric drive to keep the image where required on the plate. He just keeps a look-out for any passing clouds.

Hunt's fine eclipse photograph, Plate III, was taken with a 3-inch refractor of 40 inches focal length using an Ilford H.P.3 plate with an exposure of 3 seconds.

I will complete this chapter with a detailed account of the method I now employ to take sunspot photographs with my 4-inch refractor.

The telescope objective is a visual achromatic—not a "photo-visual", which is a special achromatic combination of lenses to bring the photographic focus and the visual focus to the same point—and it is necessary to use a yellow filter (or an orange one) with orthochromatic plates. I use plates in preference to films really because my "camera" is arranged to take $\frac{1}{4}$-plates and I have no experience with films other than with my small hand camera.

The two essential parts of my photographic equipment I consider to be the cloth-blind, spring-operated, focal plane shutter and my collection of neutral tinted filters. The camera shutter, being just in front of the plate, is not subjected to the concentrated heat of the Sun. Any shutter which operates near the eyepiece would have to be of metal, of course, or it would rapidly be destroyed.

The so-called "camera" is really just a tapered square wood box and I have to confess I do not remember where it came from. I bought it second-hand many years ago, long before I ever thought of astronomical photography, on the principle that "it may come in useful one day"—and it has. It has a thick wood front with a hole, which I enlarged to take a short piece of thin brass tube which was a fit for my telescope draw-tube. At the other end is the usual fitting for the ground-glass focusing screen and for the plate holder, in front of which the cloth-blind shutter operates. The latter can be set for different exposures up ·to 1/1,000 sec. by varying the gap in the blind, and is operated by the usual cable release.

The blind can be fixed completely open when required for centring and focusing the image on the ground glass. In other words, this is the ordinary arrangement as found in a reflex Press camera, but without the reflex part and without a lens. In fact a reflex Press camera could no doubt be adapted to fit on a telescope with the advantage that the solar image could be studied in the reflex finder right up to the moment of exposure. The disadvantage would be the weight, I should think, as such cameras are rather heavy, but perhaps the back part could be adapted to a lighter box. It is possible to buy $\frac{1}{4}$-plate cameras of this sort, without lenses of course, for a few pounds, as I have seen in the second-hand departments of camera stores, but the blind should be very carefully examined for pinholes.

My camera is sketched in Fig. 31 (a).

The other essential I mentioned was my collection of neutral filters. The brilliance of the Sun varies enormously, with the time, the season and the clarity of the atmosphere, and one must vary the light before it reaches the plate, so as to have reasonable exposure times. This is normally done by setting the camera shutter to different speeds, but, in my case, I was not too happy with the different settings as marked and I am sure that cameras rarely work at the exact speeds stated. I have therefore adopted the setting of 1/500 sec. as my standard for general use and keep my camera shutter at this. Whether it is actually 1/500 sec. does not matter as long as it remains constant. I compensate for varying solar illumination by varying my dark neutral filter which I use in combination with an ordinary x 3 yellow filter as obtainable from any photographic store. These filters are all 1-inch diameter.

Filters can be put anywhere between the incoming sunlight and the plate, but it would require larger diameter filters if placed other than near the eyepiece where the rays are narrow and, unfortunately, extremely hot. A filter must not introduce optical aberrations, of course, and this, too, is simplified and cheapened if the filter is small.

I tried various positions for the filters, cracking several with the concentrated heat, and in the end I found the position in front of the eyepiece field lens as good as any. I have made up a set of ten 1-inch diameter optically worked neutral dark filters obtained from Broadhurst Clarkson & Co. (63 Farringdon Road, London, E.C.1) for 6p each. They had a large stock of what were presumably sextant sun filters. By spending a long time on the job I was able to select a graduated number from quite light to dark. The coloured filters should be avoided. Actually the six lightest ones amply cover all my photographic requirements, but I find the others most useful for other purposes—for dropping in my eyepiece recess when observing through the diagonal, for naked-eye views and so on. The 1-inch diameter yellow filter forms the monochromatic image suitable for the

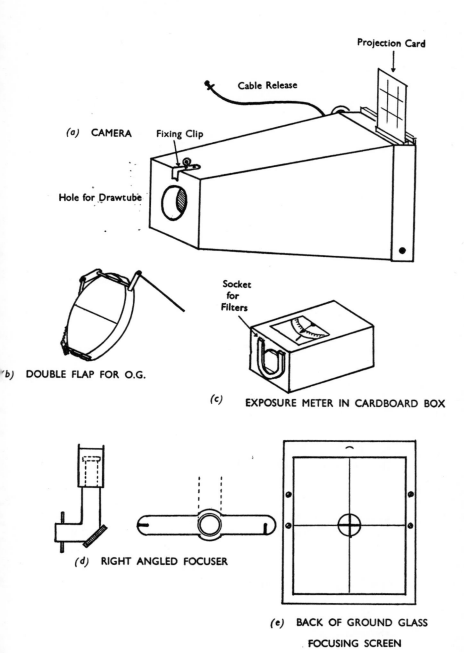

(a) CAMERA — Projection Card, Cable Release, Fixing Clip, Hole for Drawtube

(b) DOUBLE FLAP FOR O.G.

(c) EXPOSURE METER IN CARDBOARD BOX — Socket for Filters

(d) RIGHT ANGLED FOCUSER

(e) BACK OF GROUND GLASS FOCUSING SCREEN

Fig. 31. Solar camera and accessories.

orthochromatic plates. To hold the pair of filters (1 yellow, 1 neutral) I made a small brass socket to fit in the thin brass tubes of my Cooke push-in eyepieces in front of the field lenses.

Following Arber's and Wright's procedure I have recently experimented with a x 4 orange filter, in place of my x 3 yellow, but I find the results indistinguishable with my orthochromatic plates.

I have completely overcome the heat trouble by only exposing the filters to the Sun for the very brief moments of centring and focusing the image on the ground glass and for exposing the plates. This is catered for by a flap over the telescope object glass, operated from the eyepiece end by a string, as sketched in Fig. 31 (b). This two-leaf flap was made from bits of soldered brass and a spring mounted in a sweet-tin-lid rim, felt-lined to fit the telescope. My original flap was a single square of plywood, but it projected too far when open and occasionally fouled the side of the observatory shutter.

To decide the correct neutral filter to use (with my 1/500 sec. exposure) I make use of an ordinary, small, inexpensive light cell exposure meter, which I normally have for my ordinary hand camera photography. I place it in a cardboard box in one end of which is a hole cut slightly smaller than 1 inch and with slots on the outside in which I can drop two filters in front of the light cell "window". By experimenting with various filters, and wasting a few plates, I once and for all found the pointer position on the meter which would give me a correct exposure on the Sun when my camera shutter speed was set at 1/500 sec. This position was inked on the meter dial, the actual figures on the dial being ignored for this purpose. See Fig. 31 (c).

The ground-glass focusing screen has a central clear patch with a pencilled (graphite) cross, made by mounting, over the cross, a $\frac{7}{8}$-inch diameter thin cover glass with Canada balsam, as one mounts a microscope slide. If a low-power magnifier is then used on the cross the latter is seen together with the focused image of the sun. It must, however, be focused with the filters already in the eyepiece exactly as it will be photographed. See Fig. 31 (*e*).

This magnifier can be used on the ordinary part of the ground glass, but if the image is particularly steady the clear patch has an advantage, although the eye then has to be placed rather a long way behind. When the Sun is low I have the same difficulty as with the spectroscope in that I cannot fit my head and the magnifier between the focusing screen and my observatory wall, so for such occasions I made a right-angled focusing viewer out of a small brass elbow-bend tube (from "stock") by sawing off the corner at 45 degrees and fixing an oval of dark glass (cut by scissors under water) by which a dimmer image is seen by looking down the low-power magnifier to focus both the cross and the Sun. This "contraption" clips on to the back of the ground-glass frame with springy brass strips to hold it tight against the clear glass patch, as indicated in Fig. 31 (*d*). Unless the image is really steady the clear patch has no advantage, and for more general use I clip the fitting on the higher screws so that the image is seen on the ordinary ground-glass surface. I will only add as a confession that more often than not one can focus just as accurately on the screen without any magnifier at all! It was fun making the focuser anyway, and I thought it worth mentioning, as for special occasions it has an advantage. The top has a socket to take one of the neutral filters to give a comfortable view, and I would stress that when the Sun's image is "boiling" it is not only difficult to focus at all, but any attempt at photography is a waste of time.

The whole procedure is to fit the flap on the object glass end of the telescope, remove the draw-tube and, with the camera shutter fixed open, push it through the camera from the back until its flange is stopped by the front hole. The draw-tube is a tight fit and cannot easily slip, but the camera itself is locked to the telescope by the clip on the top front. The telescope adjustable counterweight is then slid along until the whole is balanced, a suitable high-power eyepiece is inserted in the 2-inch supplementary telescope and the small projection card fixed to the top back of the camera. The position of the spot or group to be photographed is noted with respect to the pencilled grid. The eyepiece is usually my 0·29 inch F/L, giving a full quadrant of a 7½-inch disk on the card. This enables me to watch a large area for turbulence and cloud.

The camera is orientated by allowing a spot to run along an E/W line pencilled on the ground glass, twisting the draw-tube in the telescope if necessary.

The chosen eyepiece (usually my 0·47 inch F/L) is inserted, without any filter, through the back of the open camera and the Sun's image centred and roughly focused. The selection of the correct filter is achieved by holding the exposure meter against the clear patch in the ground glass, the yellow filter and one of the neutral filters being in the front slot. The neutral filter is then changed, if necessary, until one is found that brings the meter pointer nearest to the inked mark. This fixes the filter combination to give me the correct exposure with 1/500 sec. and for the brilliance of the Sun as it then is.

The O.G. flap is closed, the eyepiece is removed, and, with the selected filters dropped into the field lens socket, it is replaced. The O.G. flap is opened briefly while the now dimmer and yellow image of the Sun is re-focused on the ground glass (which in my case means racking out the draw-tube by about 1 mm. only, but this is very critical), and the position of the area to be photographed checked.

The O.G. flap is then closed, the shutter set, the plate holder inserted and its cover withdrawn—all ready. This

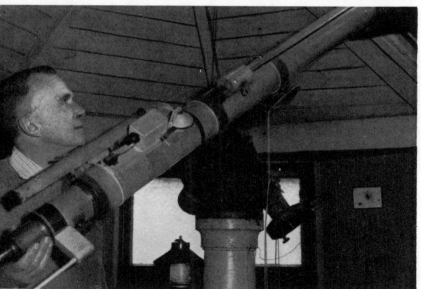

Plate IX. (*a*) The author's observatory in London, W.3.
(*b*) The author's telescope.

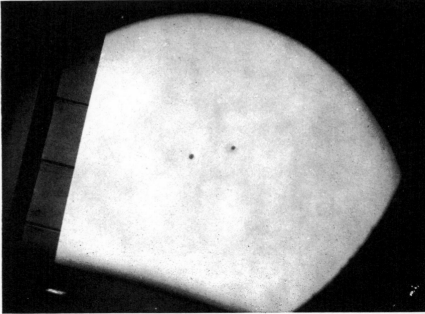

Plate X.   (*a*) Projecting the Sun's image for drawing.
　　　　　(*b*) Large projection on to observatory roof.

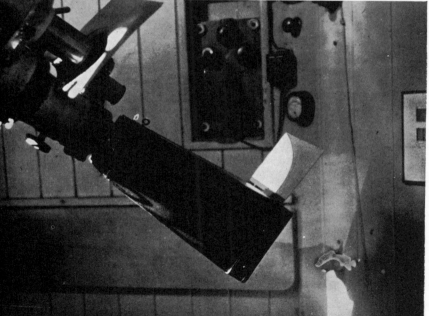

Plate XI. (*a*) Arrangement of observatory shutter
blinds.

(*b*) Camera on the author's telescope.

Plates XII-XXIV

A SELECTION OF THE AUTHOR'S SOLAR PHOTOGRAPHS

In each case the orientation of the Sun is as seen with the naked eye—north at the top and east to the left. The short horizontal lines at the bottom of the photographs represent, in each case, a distance of 10,000 miles to give the scale—the diameter of the Earth is, of course, only eight-tenths of this length.

Plate XII

Top—PARTIAL ECLIPSE OF SUN
1959 October 2, as seen from London.

Bottom—TOTAL ECLIPSE OF SUN
1961 February 15, as seen in France. This photograph
is of my domestic television screen, and it is interesting
to compare with the photograph taken direct by H. C.
Hunt from Pisa (Plate III).

The latter is a very fine photograph, but the TV
picture is, nevertheless, a tribute to the first-ever
attempt to transmit a picture of a total eclipse of the
sun by Eurovision to millions of viewers who would
not otherwise have been spectators.

The disk is a little distorted as seen on the screen and
the original image of the negative was only 0·18 inch
diameter.

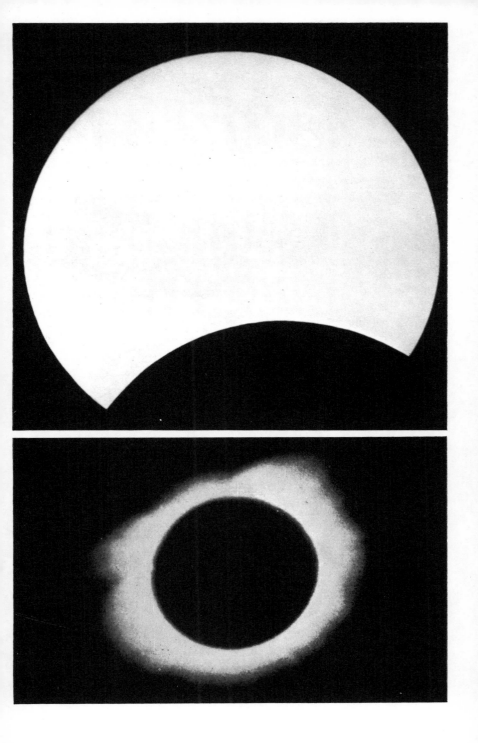

These two photographs of "regular" (symmetrical) sunspots near the Sun's limb illustrate the "Wilson Effect".

Top—1961 January 31, indicating a hollow, measured to be about 385 miles deep.

Bottom—1958 October 25, indicating a mound, measured to be about 455 miles high.

See Chapter Three.

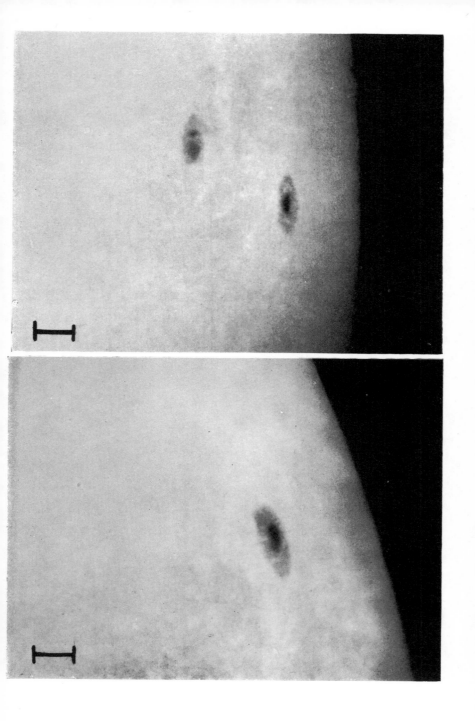

Plate XIV

Top—1958 September 18.
This is what I call a "beautiful" sunspot. Note the intricate penumbra.

Bottom—1959 January 10.
A large pair, visible to the naked eye as a group.

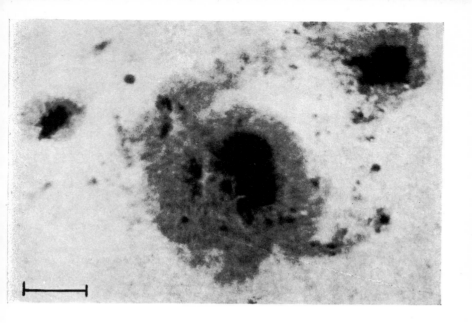

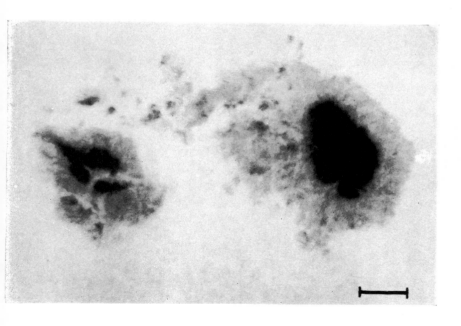

Plate XV

The top two photographs show the rotation of the sun in four days. The Sun's axis of rotation is drawn in. The bottom two photographs are of the large group and of the large spot seen in the upper photographs.

Top left—1959 August 29. Note the spot coming round the east (left) limb and the group near the central meridian.

Top right—1959 September 2. See how the large spot and the large group have been carried to the right by the sun's rotation, the spot now being near the central meridian and the group close to the west limb.

Bottom left—The large group, visible to the naked eye. It was about 90,000 miles long by 40,000 miles wide; that is, it covered an area of about 3,600 million square miles!

Bottom right—The large spot, also visible to the naked eye when near the central meridian. It is not easy to see even large spots with the naked eye when they are close to the limb.

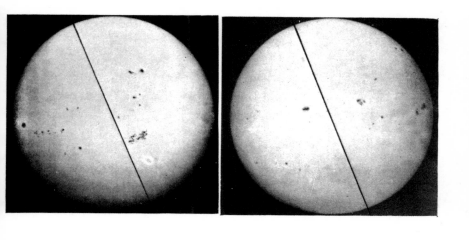

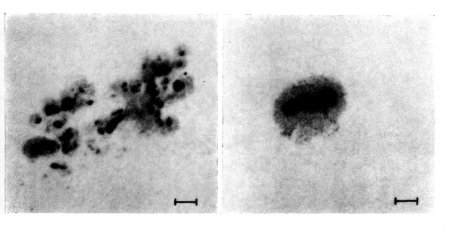

Plate XVI

Top—1959 February 19. This photograph is interesting as it shows an "island" of photosphere in the middle of the dark umbra.

Bottom—1959 November 16. Near the limb is a complicated group of spots with the white faculæ well shown against the "limb darkening". Faculæ are not readily visible near the centre of the Sun's disk.

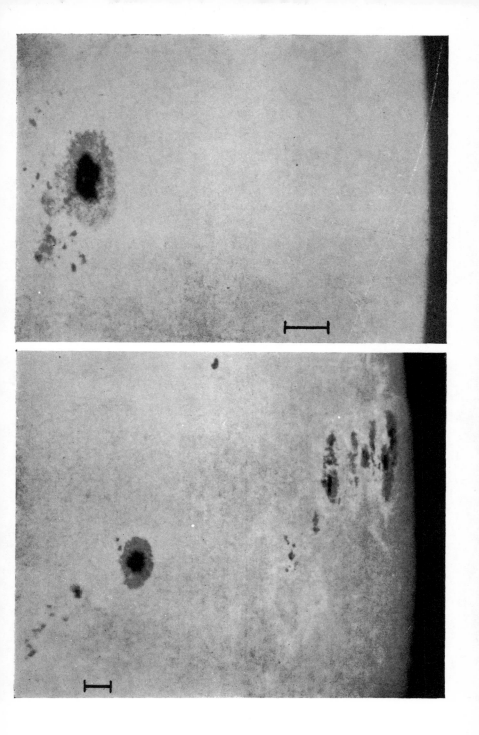

Plate XVII

These three photographs were taken of the same group at two-day intervals.

Top—1959 August 19. Shortly after the spots appeared round the east limb. Note the "bridge" across the largest spot and the complicated train of spots, and faculæ, following.

Middle—1959 August 21. Two days on and the large spot is seen more extended. The bridge is still conspicuous.

Bottom—1959 August 23. All are now getting away from the Sun's limb and the large spot is seen even more extended with its bridge closing in.

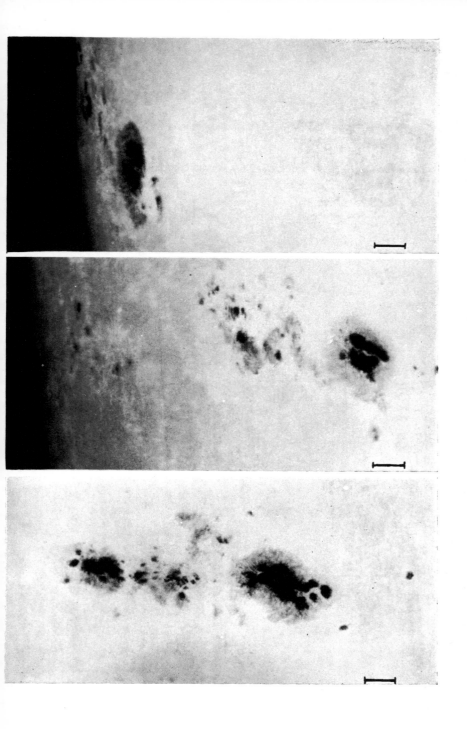

Plate XVIII

Top—1960 May 16. A large spot where a "bridge" developed.

Bottom—1960 February 8. A pair of large spots, each of which shows a "bridge" of photosphere across the dark umbra.

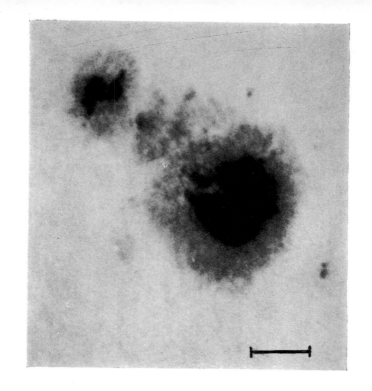

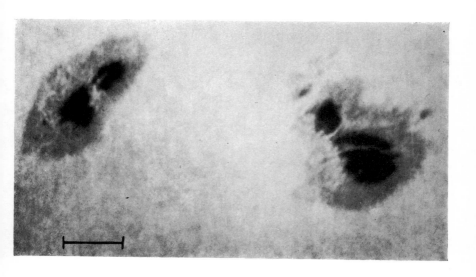

Plate XIX

Top—1959 April 4. A large naked-eye sunspot, showing "bridges".

Bottom—1959 October 16. Two large sunspots near the east limb, showing bright faculæ against limb darkening.

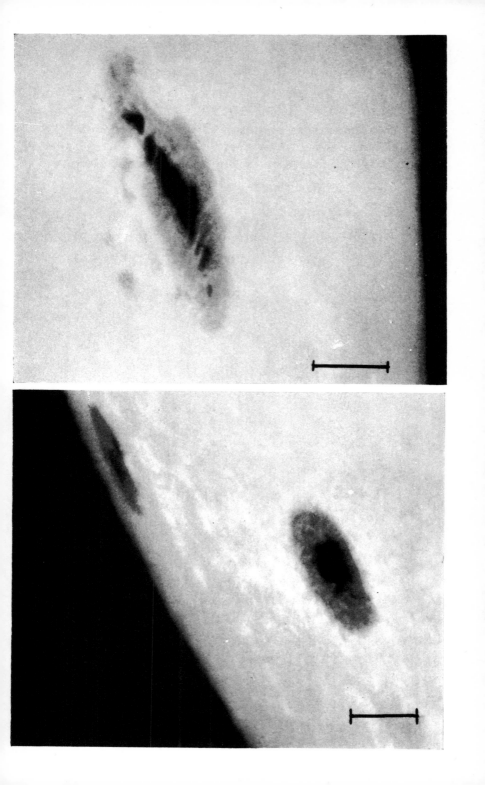

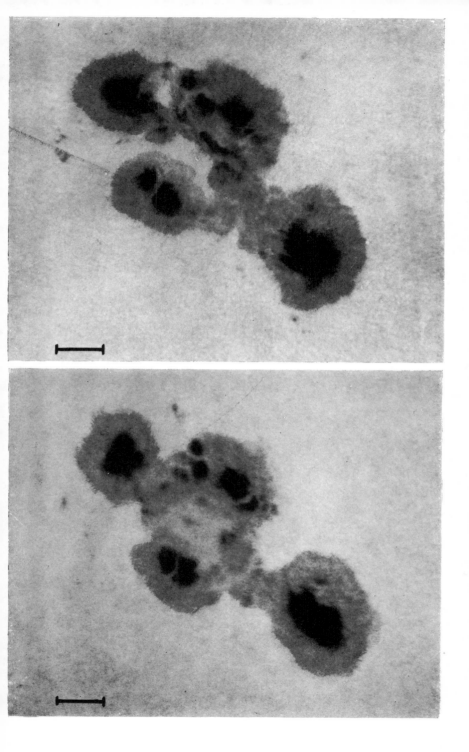

Plate XXI

Top—1959 May 10. This was a very complicated sunspot group with the individual spots taking the form of a spiral. It was a very active group which was associated with a major flare. Note the single spot near the limb showing very well the "Wilson Effect".

Bottom—1962 February 25. This was a large naked-eye sunspot group which stretched to about 100,000 miles long. The Sun's surface granulation is visible in this photograph.

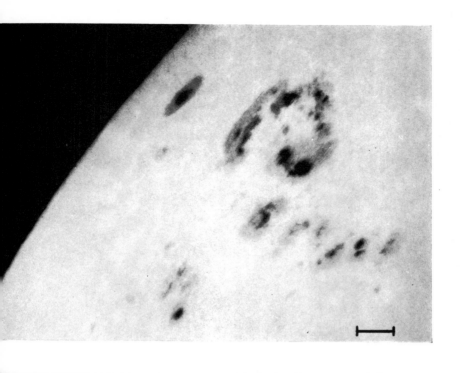

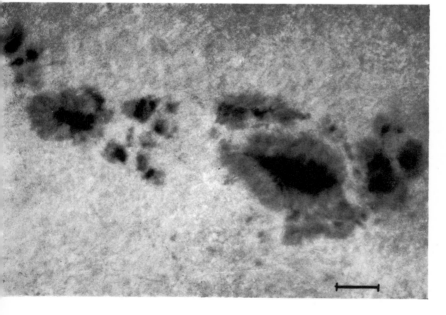

Plate XXII

1959 September 10.
A large and very striking sunspot group. The spot on the left almost gives the impression of "splashes" and the whole is quite "pictorial". This was another naked-eye group.
Surface granulation is visible.

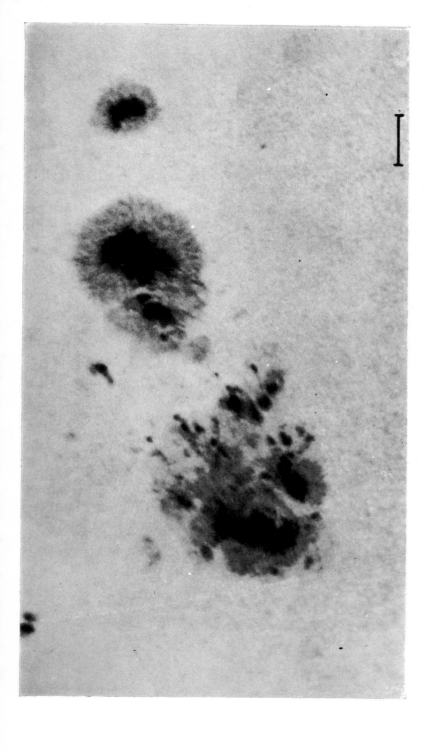

Plate XXIII

1960 November 10.
A large naked-eye spot which could again be classed as
"beautiful".
Surface granulation again visible.

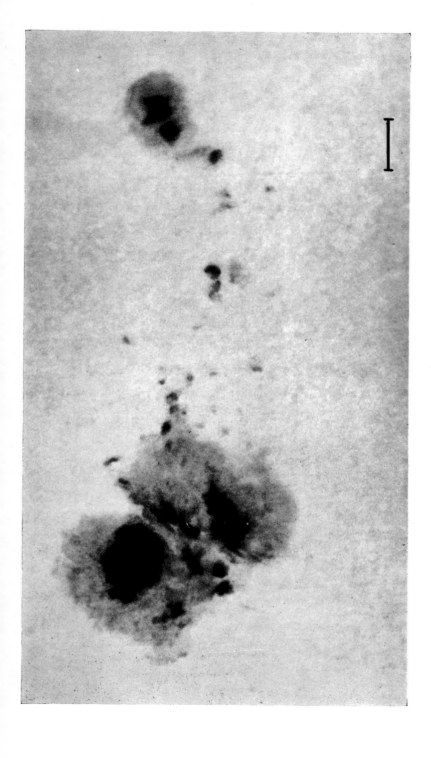

Plate XXIV

1961 July 14.
A large sunspot group with a most interesting and complicated penumbra. This group was about 60,000 miles from north to south and was visible to the naked eye. It was very active and was associated with a major flare.

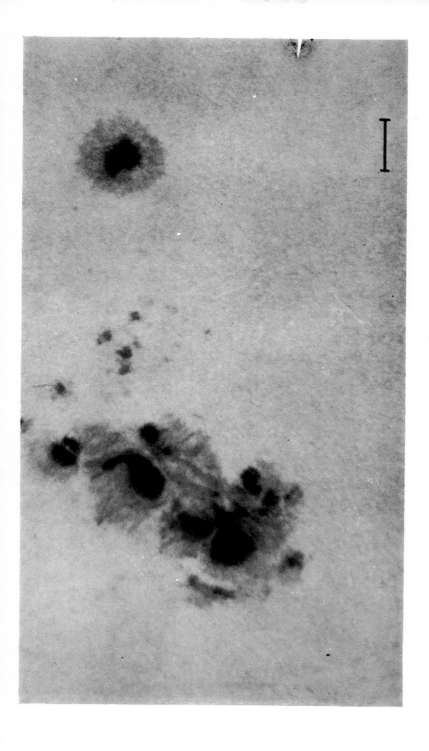

is the time to study closely the projected image above the camera, to see that the exposure is made at the most suitable moment, with no clouds passing and atmospheric turbulence at a minimum. It is here that one reaps the great benefit of having a clock-drive which keeps the solar image in place while all the preparations are going on. In any case a check must be kept on the position by the projected image relative to the grid.

At the chosen moment the O.G. flap is opened, the exposure made and the flap closed. My dark slide is a double one, and I always take a pair of pictures as experience shows that one negative is nearly always sharper than the other, due to atmospheric movements, so the operation is repeated with the second plate.

I find that immediately after the exposures the filters are warm, but not unduly so, thanks to the O.G. flap.

Plate XI (*b*) shows the camera fitted on the telescope and the projected image on the small card mounted on top.

I normally photograph individual sunspots and groups with my × 120 eyepiece, giving a solar diameter on the plate of 10 inches. I prefer this size to taking whole disk photographs as a 2¾-inch diameter image is the most one can get on a ¼-plate, but I have taken a few of these with my × 40 eyepiece. Dust on the eyepiece is a real trouble as it produces out-of-focus shadings on the picture, and therefore lenses should be kept as clean as possible with a soft lens brush.

One problem that troubled me was pinholes in the negative, and I know others have suffered in the same way. Precautions to avoid dust in the camera will naturally be taken and the plates tapped before putting them into the dark slides, and again before developing, but by examining the pinholes under a magnifier it appeared that most of my holes were not irregular, as would be caused by dust particles, but tiny circles, indicating air bubbles. I have since largely overcome the nuisance, by mixing my developer an hour or so before required, thus getting rid of most, at least, of the bubbles carried in the water.

Turning finally to the purely photographic side, I can only say again that anyone attempting solar photography should experiment and adopt the methods found to suit. But, to start with, do not think I have any special dark-room facilities; far from it, the rooms in my house not really being adaptable to this purpose. Still, improvision is all part of any hobby.

I use Ilford G 30 Chromatic, backed, ¼-plates (speed A.S.A. 10) which I insert in the double dark slide at any time of night or, if I have to do this in the daytime, in a cupboard under the stairs—which is cramped to say the least, quite apart from having to remove a lot of household "debris" and put it all back afterwards!

I develop, again in the cupboard, in Johnson's Contrast Developer (diluted 1:4) for 1½–2 minutes, with a ruby light as I like to see what is going on, and then fix in Johnson's Fixadon. When fixed, I examine the two negatives, selecting the better one and throwing the other away; sometimes I throw both away. I wash and dry in the usual way.

I enlarge to a standard 24-inch solar disk size by means of an old optical lantern for which I have made a slide carrier to hold the ¼-plates. Photographs of this disk size are convenient for calculating "Wilson Effect" measurements and are fair enlargements. My lantern is a good one optically and it now takes a 40-watt household lamp for enlarging purposes. I have a 500-watt lamp for slide projecting, but I must mention that a label still in the lantern-box lid gives instructions for trimming the oil wick!

The negative image is projected on to a vertical board with spring wires to hold the Kodak, Double Weight, White, Glossy, Hard Bromide paper (WSG3D) which I have adopted. To focus and fix the size I have scratched vertical lines 1 inch apart, with a cross between, on an old negative.

This I project on to paper marked with lines 2·4 inches apart (and other distances for larger solar disk diameters) and I move the easel (board) backwards and forwards until the projected lines coincide and are focused sharply. I again develop with the Contrast Developer and fix with Fixadon.

For lantern slides I use Kodak L 10 plates (originally $3\frac{1}{4}''$ × $3\frac{1}{4}''$ but now mostly 2″ × 2″) which are similar in action to the Bromide paper mentioned, and make contact prints.

A suitable orange light can be worked in, but without a dark room I have to wait until darkness falls, which makes a late night of it during "Summer Time". It takes me longer to set up the gear and put it all away than it does to make many enlargements, but it is all for a good cause and one is not always taking solar photographs.

The Sun's orientation on a photograph taken by the above method is the same as seen with the naked eye (see Fig. 30), but the image on the projection card is reversed mirror-wise since it is like looking at the photographic plate, in the camera, from the front.

It is useless taking a photograph when the Sun is "boiling" due to atmospheric turbulence, as already mentioned. At these times the limb is serrated and trembling all the time, with sunspots going in and out of focus as you watch them. However, it is sometimes worth a gamble if there is something very special to photograph, but a careful watch must be kept on the projection card to choose the best possible moment for the exposure.

I have explained my own methods in great detail, perhaps unnecessarily so, as I do not want to leave any questions unanswered, but my system may not suit others with different equipment. I can only say that after much experimenting in the past I have settled on the above procedures and I am now sticking to them; results must be judged by the examples at the end of this book.

We amateur solar photographers cannot compete with the wonderful and detailed pictures of the Sun recently obtained in the U.S.A. from an unmanned balloon, taken at a height of 80,000 feet above the earth, but we will continue to enjoy our hobby.

Appendix

SOLAR CO-ORDINATES (See page 127)

IT HAS BEEN SUGGESTED that an example to show the derivation of the co-ordinates would be useful to the amateur recording for the first time, and as drawing Fig. 26 (*d*) was an example taken at random from my records we will use this to derive the figures for P, B_0, L_0, for the date 1961 December 24.

The data in the annual *B.A.A. Handbooks* relating to the Sun are given for noon (U.T.) for every fourth day and the nearest date in the *Handbook* for 1961 (page 7) is December 23 with $P = +6°\cdot3$, $B_0 = -2°\cdot0$ and $L_0 = 66°\cdot0$ (last three columns).

P: This is the position angle of the Sun's axis at noon and, being positive, indicates that the N. end of the axis is to the E. of the N. point of the disk.

At noon on 1961 December 23 the N. end of the axis was inclined towards the E. at an angle of $6°\cdot3$. Since the angle at noon on December 27 (i.e. 4 days later) is given as $+4°\cdot4$ we obtain for noon on December 24, by interpolation, an angle of $+5°\cdot825$. This, to the nearest decimal which is sufficiently accurate in view of the small daily change, gives $+5°\cdot8$ and the axis can be drawn in accordingly with the aid of a 6-inch protractor.

B_0: This is the latitude of the centre of the Sun's disk, indicating the tilt towards or away from the observer. If positive, the centre of the visible disk is above the Sun's equator, indicating that the N. end of the Sun's axis is tilted towards the Earth by the angle in the table. If negative, the axis is, of course, tilted away from the Earth.

On 1961 December 23 the angle was $-2°\cdot0$ while on December 27 it was $-2°\cdot5$. Again interpolating, the angle on December 24 was $-2°\cdot125$, or to one decimal $-2°\cdot1$. This figure is entered on the drawing.

157

L_0: Here we have the longitude of the centre of the disk at the time of observation. This changes fairly rapidly and we have to take account of the actual time. As the Sun does not rotate as a solid body it was necessary for all recording purposes to fix an arbitrary (average) rate of rotation. This has been taken as 25·38 mean solar days and in the preceding page of the *Handbook* the rotation numbers throughout the year are given. December 24, 1961, fell within Rotation No. 1,448 and this is entered on the drawing.

Longitude 0° rotates with the Sun, and as the Sun rotates (as seen from the Earth) from E. to W. while the longitude on the surface is also counted from E. to W. it means that the longitude of the Sun's central meridian decreases day by day; in fact by 13°·2 per day.

For 1961 December 23 the longitude of the central meridian (L_0) at noon was 66°·0 and on noon on December 24 it must have been 66°·0 minus 13°·2 (1 day lapse) = 52°·8, but as the time of observation was 2 hours before noon we must *add* back the variation for this difference (given for various hours and minutes in the preceding page of the *Handbook*, above the rotation number) which is 1°·1 for 2 hours. 52°·8 + 1°·1 = 53°·9, which is entered on the drawing as the L_0 at the time of observation (U.T.). The resulting figures were therefore as shown in Fig. 26 (*d*).

Incidentally, 5 Active Areas were recorded for that day.

RECOMMENDED READING

GENERAL

The Amateur Astronomer, Patrick Moore (Lutterworth Press).
Exploration of the Universe, G. Abell (Reinhart & Watson).
Introduction to Astronomy, C. Payne-Gaposchkin (Eyre & Spottiswoode).

RADIO ASTRONOMY

Introduction to Radio Astronomy, R. C. Jennison (Newnes).
Radio Astronomy, F. G. Smith (Penguin).

THE SUN

The Sun, G. Abetti (Faber).
The Sun and its Influence, M. A. Ellison (Routledge and Kegan Paul).
The Sun, Patrick Moore (Frederick Muller).

Index